Country Fairs in Canada

Guy Scott

Best wishes

Guy Scott
April. 2006

Fitzhenry & Whiteside

Country Fairs in Canada
Copyright © 2006 Guy Scott

All rights reserved. No part of this book may be reproduced in any manner without the express written consent of the publisher, except in the case of brief excerpts in critical reviews and articles. All inquiries should be addressed to:

Fitzhenry and Whiteside Limited
195 Allstate Parkway
Markham, Ontario L3R 4T8

www.fitzhenry.ca

In the United States:
311 Washington Street,
Brighton, Massachusetts 02135

godwit@fitzhenry.ca

Fitzhenry & Whiteside acknowledges with thanks the Canada Council for the Arts, and the Ontario Arts Council for their support of our publishing program. We acknowledge the financial support of the Government of Canada through the Book Publishing Industry Development Program (BPIDP) for our publishing activities.

Library and Archives Canada Cataloguing in Publication

Scott, Guy, 1956-
Country fairs in Canada / Guy Scott.

Includes index.
ISBN 1-55041-121-7

1. Agricultural exhibitions—Canada—History. 2. Fairs—Canada—History.
I. Title.

S557.C3S36 2006   630'.74'71   C2006-902332-5

United States Cataloguing-in-Publication Data

Scott, Guy, 1956-
Country fairs in Canada / Guy Scott.
[232] p. : col. ill. ; cm.
Includes index.
Summary: A historical look at the evolution and the future of the country fair in Canada.
ISBN 1-55041-121-7 (pbk.)
1. Agricultural exhibitions—Canada—History. 2. Fairs—Canada—History. I. Title.
630.74/71 —dc22   S557.C3S36 2006

Cover and interior design by Fortunato Design Inc.

Excerpt on pages 189-194 reprinted with permission, Grant MacEwan,
Between the Red & Rockies, 1952, University of Toronto Press.

Printed and bound in Canada

1 3 5 7 9 10 8 6 4 2

*This work is respectfully dedicated to the multitudes
who have freely given their time, their treasures
and their hearts to agricultural fairs.*

*Their efforts have left their communities
and their country with a rich cultural legacy.*

*This work is especially dedicated to one Canadian in particular:
my late father. He truly knew how to enjoy a fair.*

*And to the thousands of Canadians who give
their time at fairs today: God bless, and carry on.*

Board of Directors, Charlton Fair, Ontario, circa 1920. Hats, dark suits and director's ribbons are proudly worn. Countless Canadians, both urban and rural dwellers, have served on fair boards over the last two hundred years. They have unselfishly volunteered their time and efforts to improve life in their communities. Wear your badges with pride: Canada and your community owe you a debt of appreciation.

Contents

Foreword vii
Introduction ix
Acknowledgements xiii

Fair Culture in Canada 1

Ancient Traditions 20

Canada's Fair History 36

The Midway: Carousels and Conmen 58

Entertainment: See It at the Fair 90

Commercial Exhibits 120

The Exhibit Hall: Pumpkins to Pictures 132

Livestock: Four Feet and Feathers 146

Parades 170

The Thrill of It All: Children and Fairs 180

Conclusion 200
Fair Listings 201
Photo Credits 209
Endnotes 211
Bibliography 213
Index 215

Foreword

THE FIRST COUNTRY FAIR I ever attended was in Shelburne, Ontario, in the fall of 1957 during the federal election campaign. I was standing in the doorway of the arena with my parents when John Diefenbaker came through with his entourage and stopped to shake hands with a small boy – me. I won a small diary in the fishpond that day and my mother still has it in her desk. On the first page it says, "Shook hands with the PM today." There is nothing else written in that diary.

The fair is one of the oldest social organizations in Canada, probably second only to the church. And there were lots of communities that had a racetrack and a fair building long before they had a church. The first fairs date back to the 1790s in the Niagara area in Ontario and they go way back before that in Quebec and the Maritimes.

I've been a volunteer at my own fair for more than a decade. I like it for the same reason I like amateur theatre. It is one of the last places where we're allowed to work together just for the fun of it. The fruits of our labour are something the whole community can enjoy … young, old, rich or poor, town or country.

The occupation of farming is in trouble these days because we have been taught to think of food as a commodity and ourselves as merely consumers. The cost of looking at food that way has been the near extinction of the family farm and the unravelling of our rural economies. We have forgotten that the objective of a food system is to sustain a human community, to knit it together and help it work and play together. One of the last reminders we have of that old system, oddly enough, is the country fair.

Here we find our connection to the land and to each other, the flowering of the competitive instinct, the pride of accomplishment and a whole lot of fun for not a great expense. It's a wonderful institution and I applaud the efforts of all those sturdy volunteers who stand at the turnstiles year after year, in fair weather or foul, to keep the tradition alive.

See you at the fairgrounds!

Dan Needles

Dan Needles, author, playwright and fair director, is best known as the genius behind the "Wingfield Farms" theatre plays.

On fair day, any and every mode of transportation was used. These three excursion steamers have brought loads of fair-goers to the Bobcaygeon, Ontario, Fair. The side-wheeled steamers were owned by the famous Boyd family of lumber fame, and plied the waters of the Kawartha lakes in Central Ontario. What better excuse for a day's outing than the local fair?

Introduction

SOME YEARS AGO, I had the honour of being on a CBC radio phone-in show whose topic was "memories of fall fairs." The producer cautioned me to be ready to fill in time between callers. He was worried this topic would not attract enough callers to fill the time slot. He need not have worried. The switchboard lit up and remained that way throughout the show. Most callers never did get on the air. He expressed surprise at such a response. I was not surprised. Fairs are an ingrained part of Canadian culture. Add nostalgia to the topic and you have a major march down memory lane for many Canadians.

The term "fair" can mean any event that presents exhibits, whether agricultural, industrial or artistic. But the focus of this book is on the first category: agricultural. While many readers will wonder how some modern fairs relate to agriculture, remember that at one time in the not-so-distant past even industrial, urban exhibitions recognized agriculture as Canada's number one industry. Agriculture may have declined in importance at many of today's exhibitions, but the basic threads still bind urban and rural shows together. Therefore, this book deals with any fairs and exhibitions that have agricultural roots.

This book uses three terms for fair: "fair," "exhibition," and "exposition." They are somewhat interchangeable.

Let there be crowds! Midway area, CNE, Toronto, 1950s. During the mid-1900s, the CNE was the largest fair in North America.

Most larger fairs call themselves exhibitions. The term "exposition" is usually reserved for huge, one-time shows such as a world's fair. But notice how the word "fair" is used for even these largest of shows. While the words "exhibition" and "exposition" give an impression of great size and grandeur, deep down they are all fairs at heart.

One of the chief concerns in creating this book was the target audience. A wide gap in familiarity with fairs no doubt exists among the Canadian public. Many readers have been faithful fair supporters, workers and attendees for years, while others are casual fair-goers. And then there is that segment of the Canadian population that is totally ignorant of the bliss of fair-going. Fair directors call this group everything from the "great unwashed" to the "great potential." Fair terminology such as "show ring," "exhibit hall," "concession" and "midway" are well-known terms to the fair enthusiast, but an unknown language to the non-fair-goer. I have earnestly tried to define and explain these terms for the novice, but I have no doubt slipped up in some instances and I apologize for some presumptions.

It was not my intention to aim this work at any one group. I hope everyone from the gnarly old fair director to the casually interested reader will understand and enjoy this snippet of Canadiana. I also hope that reading this work (or at least looking at the pictures!) will inspire the reader to think about fairs in a new light and maybe even attend one, if that is not a regular habit. Many fairs across Canada are crying out for

Concession scene, Manitoba Provincial Exhibition, Brandon, Manitoba, 1902.

attention in their own communities. They want to appeal to as many people as possible. You will never hear a fair director say that his or her fair has too many people in attendance!

When originally asked what this book was about, I replied "A history of fairs across Canada." As the work progressed, I became reluctant to use the word "history" in the title; an odd reaction from a teacher of history. But the word suggests the past, and readers are left with the impression that fairs are over with, gone or no longer exist as they once did. This is not true. Fairs are still with us, and I can safely say they will be with us for years yet. They are not a closed chapter in Canadian history; they are an ongoing chapter of Canadian culture. Yes, they have a history. Yes, they were different a century ago. But they are changing and evolving, and I would prefer to use the term "story" or "heritage," rather than "history" of fairs.

Collecting the story of fairs across Canada was an immense task: enjoyable, but far more complicated than I presumed when I naïvely began to research this subject. There are more than six hundred fairs in Canada, and each one has a story to tell. If they have not told their story, they darn well should! I read about a hundred fair histories. That should be more than enough to gather the essence of fairs, but I have that nagging feeling that there are great works with lots of relevant material I haven't found still out there … somewhere. I also viewed hundreds of photographs and I have liberally sprinkled this story with images, as I believe

it is better to let the past speak for itself. And again, I wonder: are there even better photographs somewhere out there? With this subject, I could have searched forever. I guess there comes a time when the law of dimishing returns maintains, and it is time to stop searching and start sharing.

Once, while sifting through the nearly finished manuscript, I realized that much of this book seems to centre on a particular era, namely 1900 to 1930. This is especially true of the photographs. It may leave the reader with the impression that fairs enjoyed a golden age or zenith of influence during this time span. This is not entirely true. Attendance at fairs seemed to peak in the 1950s and 1960s. But there are many fairs that attracted more people, or put on a bigger show, in the year 2000 than in the year 1900. Photography was invented by 1860 but was not in common use before 1900, so photographs from post-1900 are more plentiful. The town photographer, amateur or professional, loved to ply his craft at special local events and celebrations, and the annual fair fit perfectly. Thus the archives and collections of professional and amateur alike are crammed with images of fairs. The older photos have a certain class or import that seems absent in later eras. People dressed more formally and were conscious of posing for the camera. The subjects in those early photos have a mystical aura about them, maybe because the images no longer exist or are beyond the memory of most readers. Likewise, the old buildings have a certain magnetism, almost mysticism, due to their design. Modern structures are more utilitarian, designed to be cost effective or for multiple use, rather than ornate. The objective of this book is to present a glimpse of the past as much as to represent the fair today.

Many histories of individual fairs have been written over the years. When organizing the material for that type of history, the writer is sorely tempted to arrange the chapters by chronological dates or periods. I have chosen to group the material for this book by theme or topic. The danger of this thematic approach is that readers may lose their sense of historical chronology. This work covers a span of over two hundred years. A lot of changes have occurred within each theme or topic. For example, midways from the 1890s barely resemble their descendants of the 1990s. But the thematic approach allows the reader to study these changes. When fair-goers attend a fair or remember fairs past, they do not think about dates or eras. They see or remember shows or parts of the fair such as the midway, the exhibit hall, the grandstand show or the parade. Times and dates are more easily forgotten than individual shows or events. The thematic approach also allows us to fit different regions of the country, with their different histories and time lines, more easily into the whole story.

This work is a brief look at the "story" of agricultural fairs in Canada; a complicated and complex subject. Yet the reader must start somewhere. So, bring on the fairs!

Acknowledgements

THIS IS the toughest page of the book to write. There are so many people who have helped me in one form or another, that I fear I will leave out some names. The massive research required to compose this work involved hundreds of people from all across Canada. Despite my best efforts, I fear I will miss somebody in this acknowledgement. To these people, I apologize in advance. And now to the real heroes behind the fairs.

First, I would like to thank the "fair people" themselves. I was kindly assisted by both permanent staff from the larger exhibitions and countless volunteers from the smaller fairs. Linda Coben, archivist at the Canadian National Exhibition, supplied me with unfettered access to the extensive archives at the CNE. David Morrison and staff at the Markham (Ontario) Fair supplied many photos. Thanks are also extended to Bridgewater Fair (Nova Scotia), Paul Quinlan at Norwood (Ontario) Fair, Jim McCuaig from Metcalfe (Ontario) Fair, L. Hawthorne from Bracebridge (Ontario) Fair, Prairieland Exhibition in Saskatoon, Armstrong (British Columbia) Fair, Brigden (Ontario) Fair, Inge at Western Fair (London, Ontario), the Central Canadian Exhibition (Ottawa), Williamstown (Ontario) Fair, and finally to my home fair in Kinmount (Ontario).

Photographs came from the National Archives of Canada, Provincial Archives of Ontario, Provincial Archives of British Columbia, University of Brandon Archives, City of Saskatoon, Provincial Archives of Manitoba, Provincial Archives of Saskatchewan and especially, the Glenbow Museum in Calgary, which has a stunning collection of photographs online.

I am grateful to the CAFE (Canadian Association of Fairs and Exhibitions) for use of their fairs listings, the OAAS (Ontario Association of Agricultural Societies) for use of their materials, and for the cooperation I received from all ten provincial agricultural associations.

World's Finest Shows and Jim Conklin gave me the utmost in cooperation without hesitation.

Many individuals also readily assisted me in my research. Thank you, Elwood Jones (Peterborough, Ontario), Fred McGuiness (Brandon, Manitoba) and Ken Coates (University of Saskatchewan).

Tracey Dettman deserves thanks for seeing the project through, and especial kudos to Fortunato Aglialoro, that genius of the keyboard who made a box of papers and photos into a work of art: pure genius!

Also, a big thank you to the members of my family for enduring my obsessions when it came to the writing of this book.

Fair Culture in Canada

"It makes everybody happy."

A young boy's response to what he liked best about fairs.

MEMORIES. Memories of hot, hazy days of summer or clear, crisp days of fall. The smell of manure and frying onions. Flashing rides and towers of plush animals. The *clack-clack-clack* of the crown-and-anchor wheel. The soft bellowing of a protesting cow. The tinny call of the horse ring announcer calling for a hitch class to "reverse" or "line up!" The blare of music: full, raucous volume from the midway; softer and more deliberate from the stage entertainers. The surround-a-sound of thousands of voices: talking, laughing, murmuring. The expressions on the faces of children as new marvels unfold before their very eyes. But above all, the overwhelming sense of excitement! The immeasurable feeling that something exciting was happening deep within the bowels of that crowded assemblage known as the fairgrounds. The anticipation of delight as one enters the grounds; and the draining feeling of utter exhaustion as the fair was left behind at the end of the day. These memories of the fair seem to be strongest when we remember them with a touch of childhood's innocence. City slicker and country hick alike have fond memories of happy times spent at the fair – the sights, sounds and smells of the fair have enthralled fair-goers in Canada for generations.

A cowboy shows a practical way to pick up girls, Calgary Stampede, 1927. This is not a regular rodeo class. There was no class for girl-roping, single or double. Notice the flapper outfits. Women's fashions may change, but cowboys still dress the same!

The When and Why of Fairs in Canada

How widespread are agricultural fairs? There are roughly three thousand fairs held in North America each year. Canada itself holds nearly six hundred fairs: some large, many small. They range from the large exhibitions in Toronto, Vancouver, Calgary, Edmonton, Winnipeg, Regina and other cities to the small one-day shows in rural areas. Until recently, fairs were rated by the federal government as either class "A," "B" or "C" fairs. Class "A" fairs were the largest. This rating was based on prize money spent on accepted agricultural activities and actually had little to do with the size of the fair attendance. The rating system was officially eliminated in the 1980s and no current rating system exists today.

The Canadian Association of Fairs and Exhibitions lists fairs in Canada by province. The 2003 list gives the following statistics:

Newfoundland & Labrador 10	Ontario 230
Nova Scotia 18	Manitoba 61
New Brunswick 14	Saskatchewan 53
Prince Edward Island 15	Alberta 74
Quebec 34	British Columbia 51

When are the majority of fairs held? One "fair widow" complained there were only two seasons in Canada: winter and fair season. To some, the fair season seems to last forever. The early fairs were usually held in the fall after Labour Day. Canadian weather conditions made this schedule risky, especially in Western Canada, so the traditional "fall fair" season was lengthened. Many agricultural societies moved their dates to the summer to take advantage of nicer weather and because of other factors such as competition for midways, entertainment, exhibitors and patrons. Today the traditional summer season of July and August is the most popular

Fair-goers pose in that new-fangled contraption the automobile, circa 1904. The best place to see "what's new" was the local fair. Likely, this is the first time these venerable farmers have sat in the seat of a motor car. Like any new contraption, the motor car or "horseless carriage" was treated with scepticism and scorn by many tradition-minded individuals. Traditional horsemen were offended by the sputtering, chugging new-fangled device. The noise frightened their horses, the motor-car safety record was not good and crashes were common. Several fairs banned motor cars from their fairgrounds. Ilderton Fair in Ontario banned automobiles from their fair in 1905 and only lifted the ban in 1924! Bayfield Fair, also in Ontario, passed a bylaw banning fakirs, "refreshment booths" and automobiles in 1913. To be mentioned in the same breath as fakirs and alcohol certainly shows a lack of respect. Today, a common problem at almost all fairs is parking for the horseless carriages. The automobile, along with electricity, dramatically changed the face of the fair industry. Indeed, it also changed the face of Canadian society! So let's buckle up and take a little drive through 250 years of fair history across Canada.

time for fairs across Canada. Across Canada, 373 fairs are held between Canada Day and Labour Day. The majority of fairs in Manitoba, Saskatchewan and Alberta are held in the summer season. The fall season of September and October is the next most popular season with 170 fairs held in 2003. Most (125) of these fairs are held in Ontario. Spring (April, May and June) is the least favourite season to hold agricultural fairs, but there were still 65 spring fairs held across Canada that year. Nova Scotia, New Brunswick, Prince Edward Island and British Columbia fairs are evenly spread throughout the seasons. Labour Day weekend remains the most popular fair weekend of the year, with 52 fairs. In true Canadian style, Labour Day is a compromise weekend; the end of summer and the beginning of fall.

How many Canadians attend agricultural fairs every year? This question has bedeviled fair organizers for years. It is always a difficult procedure to count patrons at a fair. The Canadian Association of Fairs and Exhibitions estimates there were 23 million fair visits in 2001. This means the combined attendance of all fairs in Canada was 23 million. This is a formidable number. No other Canadian organization or series of events can boast such an attendance.

The largest fair in North America is the Ohio State Fair in Columbus, Ohio. In 1986, the attendance was estimated at 3.7 million people. The State Fair of Texas in Dallas attracted 2.9 million visitors. The largest fair in Canada is the Canadian National Exhibition (CNE) in Toronto. At its zenith, attendance topped 4 million. Today, it draws approximately 1.8 million fair-goers each year. It is a great achievement for any fair in North America to draw more than 1 million patrons per year. Remember, the largest of exhibitions are still only open for three weeks or less. Most of the overall Canadian attendance is racked up by small and medium-sized fairs all across the country. If you average fair attendance by the number of fair-goers, you get an average of approximately 45,000 per fair: still an impressive number. Canadian fairs are certainly more than minor events.

Canadian Agricultural Societies

Who sponsors or organizes these fairs? In Canada, most fairs are sponsored by agricultural societies, which are incorporated as non-profit organizations under the control of certain government agencies, usually the provincial ministry of agriculture. In the past, these agricultural societies were community-improvement organizations, set up to play a role in agricultural improvement and community lifestyles. This meant that early agricultural societies did more than just hold fairs. They served as livestock and machinery cooperatives. They held educational seminars and even published educational magazines. They sponsored everything from

Scenes from Conklin's Midway circa 1920. A mystical feel?

libraries to plowing matches. In the past, agricultural societies filled many roles now handled by other agencies and branches of government. Over time, the agricultural societies were reduced to holding agricultural fairs and, in many cases, operating the facilities in their fairgrounds. The fairgrounds are often the most visible reminder of the local agricultural society. In towns, cities and small communities all across Canada, the local fairgrounds now doubles as a community centre, park, sports facility and public meeting place. Gone are the livestock coops, seed sales, farm implement deals, magazine clubs, libraries and plowing matches. All these functions have been assumed by private industry or another branch of government. While this book touches on some of these other aspects of agricultural societies, such as plowing matches, seed sales and spring selling fairs, the primary goal is to tell the story of agricultural fairs.

So deeply has the image of the Canadian fair become embedded in our culture, it is often imitated by other groups. Sundry community festivals, carnivals and events often incorporate what their organizers consider the best points of the traditional agricultural fair. Basically, the local festival or carnival is cashing in on the reputation that traditional fairs have gained in Canada's cultural history. This raises the question, what is an agricultural fair?—a question that has been debated by the fair industry for ages. A hundred years ago, there were two basic criteria: a dedicated (permanent) fairgrounds and a regular date. Today, since almost all fairs are incorporated under the Department of Agriculture, certain loose guidelines have been established. It is generally accepted that to be deemed an agricultural fair, farm animals must be involved. No livestock means no fair. The involvement of livestock must be genuine: race horses and dog shows alone do not fit the agricultural bill, although these days, some livestock competitions have been replaced with animal displays or demonstrations. A second requirement for any agricultural fair is a homecraft exhibit. This can consist of sewing, baking, crafts, vegetables, flowers or field crops shown by both adult and junior exhibitors. Livestock and homecraft exhibits are the two basic building blocks for an agricultural fair. What happens after these two basics is a long story.

Fairs, Communities and Change

The history of fairs is a history of the communities they represent or serve. An agricultural society or fair is a reflection of the society of its time and place. As society or the community changes, the local fair must change with it. To the historian, the fair is a chapter in the history of the community. The decisive events, issues and changes in the local scene are all reflected in the fair. The earliest agricultural societies in Canada were livestock cooperatives and sources of education for their rural communities. They provided what the community needed: advances in agricultural sciences. As the agricultural sector evolved, so did the agricultural society. Commercial components such as new machinery, labour-saving appliances and automobiles crept into the fair. The fair became the department store for what was new in the world. Fairs in large urban areas became industrial exhibitions where agriculture was only one branch of industry. Events in Canadian and world history are vividly reflected in fair history. World wars, depressions, the invention of the automobile, use of electricity and a myriad of other events all have left, and will leave, their marks on fairs.

Changes in Canadian demographics have also had a huge effect on the fair industry. Fairs can falter or fail when their communities change. The local fair can lose its sense of identity or it must search for a new sense of community. For example, the Scarborough Fair in Ontario was cancelled in 1946 when the once-rural township of Scarborough officially became a city borough of Toronto. Gone was the traditional rural society, replaced by urban subdivisions. Unable to adapt, the once-thriving fair simply vanished. Many Prairie fairs disappeared in the 1930s as their communities were gutted by the Great Depression. Other fairs, radically affected by the decline in the rural population, had to amalgamate with their neighbours or disappeared completely. The successful fairs changed with the times and survived, even flourished. In the 1930s, as Canada was beginning to shift from a rural to an urban society, agricultural fairs also began to shift emphasis from educating the farmer to educating the urbanite. This trend contin-

Thread the Needle Race, Three Hills Fair, Alberta, circa 1910. This race was a respectable sport for ladies to participate in at a fair. It seems popular, although it will never became an Olympic sport. The attire was decidedly non-athletic.

Stampede at Hand Hills, Alberta, 1917. The stampede corral would not satisfy today's security precautions. Modern insurance agents would wince at the lack of fencing and at people sitting in front of the fence. The corrals are primitive pole enclosures, but the real significance of this photo lies in the rows of motor cars perched upon the hill. Even deep in horse country, motor cars were very common as early as 1917. The motor car revolutionized the fair industry. It boosted the mobility of the travelling public and made the distance that could be travelled in one day much greater than any horse and wagon could accomplish. Big fairs got bigger and some small ones disappeared.

ues to the present day. Countless urban dwellers get their only look at rural Canada through the local fair. This partly explains why many of the successful or growing fairs are located near large urban centres. These urban-fringe fairs are both bastions of old-fashioned Canadian rural/agricultural values, as well as theme parks, where urbanites meet the country. Fairs have become a limited-time museum, an entertainment complex and a schoolhouse all rolled into one. They provide a chance for people to learn, be entertained and reach out to their past in an atmosphere that is uniquely Canadian.

It has been said that Canada has too much geography and not enough history. As a nation, we lack common threads, common experiences. Agricultural fairs provide an experience common all across Canada. They are a thread of history found in all corners of this country, one that Canadians in every province will recognize. This page from our past has not been consigned to the history books yet. Fairs are alive and well and flourishing. They may have changed over the years, but they still explode upon their community scene at least once every year.

One of the premises of this book is that agricultural fairs are both part of, and influenced by, Canadian history. To understand the changing roles of fairs, it is necessary to have a rudimentary knowledge of Canadian history; or at least of the major events that shaped this nation. Fairs were common throughout Europe for centuries before they were transplanted to North America, primarily by British settlers. The formation of agricultural societies followed on the heels of agricultural

Indian encampment, Calgary Exhibition and Stampede. Most Western Canadian fairs feature participation by Native peoples. Fairs are community efforts and successful fairs involve the whole community. Native peoples enjoyed showing off their culture at that local cultural event, the agricultural fair.

settlement. When Nova Scotia held its first fair in 1765, New Brunswick and Prince Edward Island were unsettled, Newfoundland a mere fishing station, Quebec only two years under British rule, and the rest of Canada virtually unexplored. Agricultural societies were quickly established in all settled regions after the British took control. The establishment of these societies was government policy, and agricultural societies were given grants to back up this policy and ensure their survival. When Western Canada (Canada west of Ontario) was acquired after 1869, settlement and agricultural societies proliferated throughout the area almost simultaneously. The first three institutions established in a new settlement were a school, a church and an agricultural society; and not necessarily in that order! Prominent individuals felt it was a necessary duty to be on the board of directors of the local agricultural society. Any biographical sketch of a prominent person in the nineteenth century will proudly proclaim him to be a member/director of an agricultural society. This title garnered the same respect as being a prominent church member, lodge member or member of a political party. Political office, whether national or local, demanded an affiliation with an agricultural society, even if it was honorary. Businessmen, lawyers and just about everyone who was not directly affiliated with agriculture felt it was incumbent upon them to belong in some capacity to a local agricultural society. The founding board of directors of the Miramichi Fair in New Brunswick consisted of twenty men from the following professions: Merchant – 7, Hotelier – 3, Barrister – 2, Accountant – 1, Clerk – 1, Agent – 1, Millwright – 1, Manufacturer – 1, Liveryman – 1, and Farmer – 2. The agricultural community was certainly not overrepresented! Clearly, the board of directors was composed of the most prominent men in town, and not just farmers.

The pageant *Montezuma*, CNE, 1933. *Montezuma* was the curtain call for noted Canadian producer George Penson. The production takes certain liberties with historical fact, but most fair-goers didn't seem to mind. Penson coordinated pageants at the CNE from 1894 to 1933.

The earliest Canadian fairs were almost exclusively agricultural. Canada in the 1700s and 1800s was an agricultural nation. Even fairs that dubbed themselves industrial exhibitions celebrated agriculture as the number one industry. Only in the last half of the twentieth century did agriculture lose its stranglehold on some fairs. Today, most Canadians have no direct ties to the farm sector. Less than ten per cent of Canadians live on a farm. Yet the system of agricultural fairs marches on, majestically unabated, altered but not diminished.

Fairs have adapted to changing times. They are more than agriculture; but then, they always were. Yet the more they changed, the more they stayed the same. The fair that totally forgot its roots was doomed. This balance between old and new, rural and urban, agriculture and industry, education and entertainment, is a fine line that any fair still alive has had to walk throughout its history. The twenty-first century fair is a mix of nostalgia and space-age technology.

One of the best ways to study changes in fairs is through their names or titles. As times and rules change, so do the names of the fairs. The Calgary Fair became the Calgary Stampede, Edmonton now holds Klondike Days, and the Toronto Industrial Exhibition morphed itself into the Canadian National Exhibition. The East Surrey Agricultural Fair became the Cloverdale Rodeo, and the Vancouver Industrial Exhibition adopted the title Pacific National Exhibition to reflect a changing role in its community. Regina now holds Buffalo Days, while Winnipeg sponsors the Red River Exhibition. The adjective "Royal" was injected into the Toronto Winter Fair to emphasize its importance and to reflect the fact that exhibitors come from all across North America to participate. Other fairs changed their names to reflect changing geography and history. The East Electoral Division of York County Agricultural Fair in Ontario shortened its title to Markham Fair. The Galway and Somerville Townships Agricultural Fair was changed to Kinmount Fair to make its identity more geographically simple. Many provincial governments have encouraged fairs to change their names to fit current political and geographical realities. Few Canadians today can identify townships or counties outside their own area.

To the dedicated fair-goer, fairs have a culture and subcultures all their own. The subcultures are many and varied. Some obvious examples of subcultures are the midway, livestock shows, exhibit hall, truck and tractor pulls and demolition derbies. Each subculture has its own vocabulary, traditions, tricks of the trade and sense of camaraderie. Some subcultures are permanent lifestyles for the people who belong to them, such as the midway workers. Some are temporary diversions from another world, for example livestock exhibitors and tractor pullers. To the countless armada of people who organize fairs, fair subcultures are lived for only a few days each year. For the duration of the event, fairs become villages or towns of their own. They quickly develop a sense of community, sharing common interests and goals. There is a sense of family within the fairgrounds, and a friendliness develops among disparate groups. The common goal of staging a successful fair drives these disparate groups to cooperate beyond normal expectations. The village theme can be readily noticed in most fairgrounds. They have their own boundaries, streets, entrances, directional signs and districts; for example, the midway, livestock barns and exhibit hall. Fairs offer maps showing their component parts, and many even put up street signs or directional signs within the town limits. They have their own electricians, plumbers, carpenters, police, parking and even their own hierarchies. They become a community of their own for the few days of the fair.

A fire brigade and station, CNE, early 1900s. The CNE was a city unto itself for the duration of its schedule. It had its own streets, services and hierarchies. It even had its own fire department! The insurance companies must have loved it.

Every fair is held together by its legion of volunteers. They have always been the backbone of Canadian fairs. In the 2001 International Year of the Volunteer, no organization stood as tall as the fairs. To put it bluntly: no volunteers, no fairs. Why do people volunteer at the fair? There are two main reasons. The first is a sense of civic duty. People want their community fair to keep going, be successful, survive for the coming generation. As one director said, "If I don't organize the horse show, it might disappear." It is a chance for citizens to give back to their community, to make it a better place. A second reason for volunteering is a sense of self-worth: to be part of something successful makes the volunteer feel

Maurice Ingeveld, a rancher recently emigrated from Belgium, poses in his shack near Millarville, Alberta, spring 1908. His worldly possessions are clustered close to his bed. Notice the poster of the Dominion Exhibition to be held later in the year at Calgary. Fairs were considered to be great assimilators of new immigrants. For new Canadians or old, the agricultural fair was an essential element in Canadian culture.

Poster advertising the Dominion Exhibition at Calgary. Even as early as 1908, Westerners were worried about the "vanishing lifestyles" of the Prairies. Banff Resort also gets a plug. The slogan "another trail cut off" seems to ominously predict big changes on the Alberta scene. Actually, cowboys are still a prominent part of the Stampede today, and have not totally disappeared from the area.

good. "It makes you proud to be part of something so good," was how one volunteer summed up his reason for helping at his local fair. People like to be associated with successful ventures. Many volunteers join the fair out of a sense of civic duty, but stay on the job for the feeling of self-worth. If the volunteers were ever to lose their sense of civic duty and their self-satisfaction, the fair would be in dire straits.

The Appeal of the Fair

Canadians attend fairs for many reasons. Exhibitors can show off their accomplishments: whether they have a special calf, a favourite horse, a beautiful quilt or a giant pumpkin, they get a chance to show off their pride and joy to the community and be rewarded for a job well done. It is also an opportunity for exhibitors to compare their best efforts with the work of others. Your pumpkin may be huge, but is it the biggest? That question can only be answered at the fair. The competition is gentle and friendly. No losers were ever executed or fired after the fair. Very few exhibitors show for the prize money. With most exhibitors, their smiles speak volumes about why people show at a fair. And if you don't win, there's always next year! Livestock exhibitors can use the fair to promote their business. Buyers and sellers mingle freely and informally in a relaxed setting. Other groups use the fair to promote and educate. The educational component is often subtly presented, but learning is still an important facet of every fair.

Sawyer and Massey 13 H.P. portable steam engine, 1898, North-West Territories. Equipment manufacturers and their local dealers made their living by advertising at the fairs. Almost every farmer in the district would be at the fair, making it a perfect place to sell or demonstrate a new line of machinery. Very few farmers bought on the spot, but sales after the fair were much improved by having a display at the local exhibition. The Massey Machinery Co. made itself famous by participating actively at local fairs. A letter to the *British American Cultivator*, an early farm magazine, proves the value of advertising at the fair. One reader wrote: "I saw at the fair held in Rochester (1864), an agricultural boiler for steaming food for cattle. What became of it? I would like very much to obtain such an apparatus if it answers a good purpose." [1]

Fairs are also social meeting places. They provide a chance to meet old friends and make new ones. Fairs have always cashed in on this social theme. Many former residents come home to meet old friends and relive old memories, at least for a day. This homecoming was often given a special day at the fair and labelled "Old-Timers Day," "Homecoming Day," or "Citizens Day." Fairs can also showcase subtle regional differences. While this book intends to show that fairs across Canada have a lot in common, they also have their regional distinctions. Fairs in Western Canada are big on rodeos, British Columbia fairs show off their timber specialties, and East Coast shows usually feature some sort of maritime theme. And even among provinces, specialties occur: fruit displays in Southern Ontario, dairy shows in Quebec. Each community appeals to local pride by putting forth its best or most famous features.

Escapism has always been a big part of fairs. The fair is a different world, a culture all its own. The sights, sounds and smells of the fair cannot be matched by television or computers or books. The atmosphere of a fairgrounds is mysterious, enticing and friendly all at the same time. The midway, for example, has a medieval feel: it reeks of mystery, romance and excitement, all set among flashing neon lights and blaring music. The fair becomes literally another world, a transient village, a once-a-year Brigadoon or Shangri-La. In the modern era, with the mysteries and marvels of the world just a click of a button away, the mystery and mystique of the fair is harder to maintain. Yet somehow, the mystique of the fair remains.

But the key reason most people attend fairs is to have fun. The fair is designed for people to enjoy themselves. While much has been said about a sense of duty, showing one's accomplishments, learning, marketing, meeting people and escapism, without the fun element, fairs would never exist. Fun and good times are driving forces behind all fairs. If fair-goers do not have fun or enjoy their fair visit, they will stop attending and the fair will cease to exist. A famous story best sums this up. One day, the manager of a large fair was watching the crowds thronging the fairgrounds. He mused to himself, "Why do they come?" He began to walk through the crowd, asking patrons what they liked best about this fair. He stopped a little boy and asked this question of him. He expected a reply like the midway or the games or the entertainment or even the animals. Instead, the little fair-goer thought carefully for a moment and bluntly declared, "It makes everybody happy." This lesson has not been lost on the fair industry.

Diving horse, 1913. If women could put on diving shows, why not the noble horse? Both horse and rider show fine form as they plummet into the special water pit. An equally impressive feat was getting the horse up a special ramp to the top of the jump!

FAIR ORGANIZATIONS

Every year fair directors meet at a convention to decide what entertainment and programs will take place at the upcoming fair. Most fairs across Canada send directors to these conventions to buy entertainment, as well as talk over issues related to fairs and do a little networking. Fred McGuinness, a director of Brandon Fair, described how entertainment was booked at a convention.

"Each year, when the first performance kicks off another week of night shows, there were two or three of us who had previously seen most of the acts. We were members of the exhibitions attractions committee. It was part of our duty to visit Chicago in December during the annual meeting of the Showmen's League of America. Each evening there was a banquet which was followed by a parade of talent, performers whom their promoters viewed as ready to venture out on to the fair and exhibition circuit.

"Only once each year would this audience be gathered in one hall. Included were directors from all fairs in North America. In the wings were persons who referred to themselves as 'impresarios,' but were known to the trade as 'flesh peddlers.' They were the booking agents who would provide work for the entertainers. If Fred Kressman, say, or Ernie Young, could assemble a two hour show for $10,000 a week, then he would set about finding bookings for which he might be paid twice that amount.

"Over the many years of the 'A' circuit on the Prairies, an elaborate means of selection evolved. Late on the morning of the last day of the Showmen's Convention, we members of all the attractions committees on this circuit gathered in a committee room in the Hotel Sherman. The chairman duly cautioned us that no one could leave the room until the final votes were counted. Lunch was served inside. If someone had to go to the rest room, an attendant went along so that no surreptitious telephone call could be made to tip one producer on to what his opposition was offering.

"Those impresarios bidding for the six weeks circuit, the 'A' circuit plus the Lakehead Show, drew lots to make their presentations before the committee. One after another they outlined their theme, explained its significance, then began the detailed pedigrees of

Hudson's Bay Company fort replica at Calgary Stampede. A slice of Canadian culture and history at the fair.

those acts they held under contract. I can recall the whole lot of us writing feverishly as we listed the names of jugglers, acrobats, balancers, vocalists, and high-wire performers. We soon learned to place heavy discounts on words like 'smash,' 'boffo,' 'socko' and phrases like 'show stopper.'[1]

"For the uninitiated, Showmen's League meetings were a learning experience, and one of the most important lessons was that there is a complete society under the generic name of 'show business.'"

Ontario had formed a fairs association as early as 1846. While this group mainly organized the provincial exhibition, it also coordinated fair issues and dates. By the early 1900s, Ontario fairs were divided into fifteen regional districts. Since nearly 400 fairs were held annually in the province, conflicts over dates were numerous. Almost all these fairs clustered their shows in the six weeks between Labour Day and Thanksgiving. Competition for midways, exhibitors and entertainment became intense. With the ease of transport created by motor vehicles and better roads, this competition became even more fierce. As a result, many fairs began to choose new dates, especially ones in July and August. The fair season was extended. Eventually there was at least one fair every week between June and Thanksgiving.

Western Canada suffered through a hasty and haphazard growth of fairs between 1885 and 1914. Soon fairs found themselves competing for everything: stage shows, exhibitors, midways, judges, patrons. As early as 1905, Manitoba organized a "fair calendar" to allow better use of department judges. Regina and Saskatoon bickered over fair dates in 1909. Both wanted to follow Winnipeg and Brandon. In 1911, delegates from the three Prairie provinces formed the Western Fairs Association. By mutual consent, they hammered out a circuit where fairs organized their dates to avoid conflict with one another. There was to be a class "A" circuit of Calgary, Edmonton, Saskatoon, Regina, Brandon, Winnipeg, Vancouver and Victoria. A second group of class "B" fairs also organized a series of circuits, while the smaller "C" fairs were left pretty much on their own. A class "B" fair had to offer a minimum of $4,000 in prize money and take in $3,000 in gate receipts.

This new organization promptly passed five resolutions:

1. Fair dates to be set in advance at a meeting.
2. Dominion and provincial exhibitions to be cancelled and the money divided into annual grants to the agricultural societies.
3. Each fair to adopt the same colour of ribbons.
4. A "pass policy" to be implemented consistently among fairs.
5. Fairs to bargain jointly for midways and stage shows.

The class "A" circuit quickly became a prize plum for the big midway operators from the US. It guaranteed them at least five major exhibitions in a convenient circuit. At the Showmen's Convention, usually held in Chicago, the class "A" circuit negotiated with producers each year for a suitable show. In 1914, F. Barnes won the grandstand contract for $1,500 per fair.

Early in their history, Prairie fair organizers began to realize they had a problem with dates. The earliest fairs were scheduled for September and October, to match their counterparts back in "old Ontario." However, Prairie weather patterns are decidedly different from Eastern Canada's. After a few snow-outs, most Prairie fairs changed their dates to July and August to take advantage of better weather conditions.

In an effort to improve the availability of midways and entertainment, the North Pacific Fairs Association was formed in British Columbia in 1916. The member fairs dealt with three issues of immediate concern:

The midway at night. A large midway laid out in standard fashion. Electric lighting certainly changed the midway operations.

1. Standard midway contracts were drawn up.
2. Recommended lists of entertainers and carnival companies were issued. Likewise, a blacklist of the same was compiled.
3. A schedule of race dates and the horse registry for the same was compiled.[2]

British Columbia had already organized a fairs association to handle fair dates and schedules by this time. The province was divided into five regions, and dates were agreed to by regional fairs to avoid conflicts. The Lower Mainland Fair circuit for 1910 was as follows:

Vancouver [PNE], August 15 to 20
North Vancouver, September 9 and 10
Central Park Vancouver, September 21 and 22
Delta [Ladner], September 23 to 24
Surrey [Cloverdale], September 27
Langley, September 28
New Westminster, October 4 to 8
Richmond, October 11 and 12

Vancouver, being a new addition, chose August for its dates. This broke the cycle of September and October fairs, but Vancouver made no bones about being a purely agricultural fair. It called itself an industrial exhibition and set its dates to avoid other fairs and cash in on the summer weather.

Today, all Canadian fairs belong to a provincial organization. Each province has an Association of Agricultural Societies, a body to regulate the fair industry. These central bodies coordinate the activities and operations of societies and fairs in their jurisdictions. This includes holding conventions and regional meetings to talk about issues affecting fairs.

Historically, despite the excellent work done by the provincial bodies, it was felt that an all-Canadian organization of fairs was needed. Thus, in 1924, thirty-six delegates representing sixteen fairs from across Canada met in Toronto to found the Canadian Association of Fairs and Exhibitions (CAFE). The all-Canada organization has grown over the years, until today it works on behalf of the nearly six hundred agricultural fairs and exhibitions across the country. The CAFE holds a yearly convention and publishes a magazine called *Fair Scope,* dealing with fair issues. The CAFE also lobbies various governments and fair-related organizations on behalf of Canadian fairs. Every year, the CAFE collects the fair dates and addresses of all agricultural fairs in Canada.

Two other organizations are also very important to the fair industry. The International Association of Fairs and Exhibitions (IAFE) represents fairs and exhibitions from all across North America. The IAFE holds an annual convention, usually in Las Vegas, that allows fairs from all over the continent to meet, discuss problems, network and view a huge trade show. Long associated with the fair industry has been the Showmen's League of America. It was founded in 1912 by people involved in the midway/carnival/entertainment industry. The first president was Buffalo Bill Cody, that famous American showman who formed the *Wild West Show,* a touring show of Cowboy and Indian acts credited with being the most famous touring-show act of the early 1900s (with apologies to P.T. Barnum). The Showmen's League was based in Chicago, where it still maintains its headquarters. The yearly convention became a prime event for larger fairs and fair organizations to attend every year. Here, every entertainment under the sun could be booked for the fair season. In 1959, showmen from Canada founded the Canadian Chapter of the Showmen's League of America based in Toronto. Every year they sponsor a convention in conjunction with the Ontario Association of Agricultural Societies' annual convention. After all, fairs and carnivals were always closely related!

Bulkley Valley Fair, Telkwa, B.C., 1912. When permanent buildings were not available, crude tents were substituted. This scene was shot from a very early stage in the fair's history. As the fair matured, better facilities were erected.

Family picnic at Erin Fair, Ontario, 1910. The local fair was an important holiday for the farm community. Part of the fair was a chance to meet family and friends for a picnic lunch. All ages could enjoy a fair. The family conveyances are parked in the background. Attire is very formal, likely their Sunday best.

THE FAIRGROUNDS: FOCAL POINT IN THE COMMUNITY

In their formative years, most fairs did not have permanent fairgrounds. They rotated between major centres and usually squatted on a temporary site. These sites could be public holdings like the town commons, a street, an empty lot, or private properties like a hall, company yard or private farm. In the earliest days of the Saanich Fair in British Columbia, it was customary for the president to hold the annual exhibition in his barn and fields. Hosting the fair went with the office until a permanent fairgrounds was acquired. In larger towns and cities, public parks were an obvious site. Tents were liberally used to shield exhibits from that curse of all fairs, rain. Many fairs were nicknamed tent cities, a term still used today. Livestock and horses were tethered to temporary hitching posts or fences. Pens were often built the day of the fair to hold small animals, but some pigs and sheep were never unloaded, being judged right in the wagons in which they had arrived.

These primitive conditions were not satisfactory in the long run. The ability to use the same site year after year was often uncertain due to property-ownership changes and other factors. Borrowed fairgrounds were often unfenced, and it was difficult to collect admissions or control crowds if the site was unfenced. While tents are always a big part of fairs, permanent buildings are more desirable. It was always prudent to build on your own land. Since most fairs featured some sort of horse racing, a racetrack and a grandstand were also priorities. Therefore, the ultimate objective of any respectable fair was to have its own fairgrounds. Fairs could change sites frequently over their early history, but once buildings began to replace tents, moving locations became unsatisfactory. Thus, most fairs became rooted in their locations after an initial pioneer period.

Battle River Fair and Stampede, Hardisty, Alberta, circa 1920. A good example of a temporary fairgrounds. Livestock pens and show ring (without fencing) are in the little valley. The tent city contains concessions, displays, exhibits and other trappings of the fair. Lots of room for parking, picnicking, etc. The next week, this site would be an empty pasture. Such was the nature of many early fairgrounds.

It took many years for a fair to build up its facilities, but it could be done. The South Wellington Electoral Division Fair at Guelph, Ontario, showed a good example of the growth of a fairgrounds. It began in the 1850s in a borrowed field. By 1871, it boasted the following facilities:

> 33 acres surrounded by an eight-foot board fence
> four buildings: a large exhibit hall, an implements hall, a grain/produce exhibit hall, a poultry building
> 600 ft. of horse stalls
> 900 ft. of cattle barns
> 500 ft. of sheep and pig pens
> 400-ft. horse-judging ring
> 3 livestock show rings [1]

The only items missing were a grandstand and a racetrack from this list. The stalls, barns and pens were not enclosed buildings, but the open pole-barn type: long, narrow, opened on the sides with the roof covering the actual stalls. Some of the larger and later-in-the-season fairs eventually went to completely enclosed buildings, but the pole barn and outside show ring remain popular to this day.

Exhibit halls took many shapes and sizes. They ranged from simple sheds to elaborate crystal palaces. Fairs were, and still are, often judged on their facilities, particularly their buildings. The more elaborate or spectacular the buildings, the larger and more affluent the fair. The histories of individual fairs are filled with references to buildings constructed, demolished, burnt and replaced. The minutes of many agricultural societies mention repeatedly the need for new buildings, better structures to replace the old ones and the desire to improve the facilities on the fairgrounds.

As time marched on, the fairgrounds changed. As winter sports such as curling, skating and hockey grew in importance, large covered arenas became necessary. These arenas were often built in the local fairgrounds so they could serve many purposes. They could be used for winter sports in the off-season, and during the fair, they could be utilized as exhibit/display halls or livestock buildings. The fairgrounds in many communities was the main, or only, public park of any size and hence the logical site for arenas and other sports facilities. Many of the original arenas were joint enterprises between the agricultural society and local government or service organizations. By the late 1900s, many societies withdrew from the arena management business due to increasing cost and complexity of operations and management. While the arenas often remained on the old, original fairgrounds, local municipalities assumed their ownership and management.

In the last half of the twentieth century, fairgrounds have undergone many changes. In larger towns, urban sprawl often enveloped the grounds, as rural pastoral settings gave way to subdivisions, malls and high-rises. The fairs were squeezed for space and this often necessitated a move to greener pastures beyond the urban sprawl. Other fairgrounds became even more multipurpose. The fair facilities catered to the demands of their community. Non-agricultural events such as sports, trade shows, horse racing and off-season events all gravitated to the local fairgrounds. In some larger centres such as Toronto, Calgary, Edmonton, Vancouver and Ottawa, the local fairgrounds became the centre for sports, trade shows and special events. The grounds often had the best facilities, parking, easy access and the local tradition of being "the place" for such events.

As cities sprawled, the local landscape and community changed. Fairs, even industrial exhibitions, became irrelevant in many urban centres. Cities such as Hamilton, Montreal, Windsor, Kitchener and St. Catharines cancelled their exhibitions and liquidated the fairgrounds. Many smaller local fairs also disappeared. The main reasons were the decline in the rural population, economic stresses such as the Great Depression, and changes within the community. The advent of the automobile did much to change the face of fairs by making Canadians more mobile than ever before. They were no longer limited by transportation difficulties to the nearest fair. Now they could voyage long distances to take in fairs of their choice.

As a result of these many factors, some fairs gave up control or ownership of their fairgrounds. In most cases, local governments assumed control because it was just too complicated or expensive for the local volunteer agricultural society to operate the fairgrounds. Often the sites were surrendered in exchange for free usage by the agricultural society during the fair time. Surrendering the grounds put many fairs at the mercy of local politics. The fair board versus local government is a common situation in many communities all across Canada. Larger centres were particularly prone to political meltdowns. It was not uncommon for the board of directors to be fired, or to resign, in conflicts with local governments. For example, the directors of the Pacific National Exhibition (PNE) in Vancouver were replaced en masse several times after disputes with the city council. The disagreements always seem to hinge on usage of the fairgrounds. In almost all cases, the show still went on … in one form or another. A good example of local politics vs. the fair board occurred in Vancouver in the 1970s. A new provincial government fired the board of directors of the PNE because certain groups in the community had dissatisfaction about issues such as traffic, use of the fairgrounds, noise, planning, etc. The new provincial government seized on these ill feelings and replaced the board with a new group meant to represent the "new Vancouver." When asked about the future of the grounds, the new fair president stated: "You bet your sweet bippy we wanted to change it [the fairgrounds] into a giant community centre." Eighteen months later, another new provincial government went back to the old system. The fair and other activities still went on regardless of these changes. After all politics are politics; and fairs are fairs.[2]

Fairgrounds are still very much a part of the landscape in many Canadian communities today. This is especially true in the small rural communities, where they are still considered a public park. In the larger and medium-sized communities, the local fairgrounds is often a large, multipurpose facility used year-round for a dizzying array of events. While the fairgrounds may have changed their appearance and their function over time, they still retain their role of serving their communities as a public place.

Ancient Traditions

FAIRS ARE ALMOST AS OLD AS RECORDED HISTORY. Ancient fairs served two basic purposes. First and foremost, they were market fairs where goods were bought and sold. In the ages before malls and department stores, fairs were commercial marketplaces. The second purpose was the carnival or festival atmosphere. While the commercial side of the fair was the official reason for their existence, the entertainment side also played a big part. Fairs were holidays, social gatherings, escapes from the everyday world and a time of fun and pleasure.

Ye Olde Trade Fair

The Bible makes references to the fairs of Tyre — early trade fairs in Tyre, the leading commercial city in Palestine. The ancient Greeks were also famous traders and held trade fairs for marketing sundry goods. This tradition was passed on to the Roman Empire and hence into Western Europe. The trade fair survived the fall of the Roman Empire and continued to be a vital part of whatever commercial activity arose from the ashes of that great empire. Charlemagne, the first great emperor of the Holy Roman Empire (circa 800) encouraged trade fairs as part of a policy of reviving commerce in his empire. Alfred the Great, king of England (circa 900), was also a promoter of commerce and the trade fair. As economic activity and travel revived in the Middle Ages, so did the need and demand for trade fairs.

In the Middle Ages (circa AD 1000–1500), many great trade fairs flourished all over Europe. In Russia, Viking entrepreneurs established the great Nizhni-Novgorod fair. At this important commercial city, East met West for the mutual exchange of goods. The great fair at Troyes in France began the practice of weighing money to avoid the use of underweight, or "clipped," coins. Thus the term "troy weight" for measuring valuable metals. Flanders, in what is now modern Belgium, became a great cloth-producing region during the Middle Ages. To promote the cloth trade, large market fairs sprang up all over Belgium and France. Flemish merchants also wandered abroad to sell their goods, giving rise to countless other fairs in England, France, Germany and Northern Europe. Other merchants followed the same fair circuit, all dealing in exotic goods such as wine, metalwork, spices, furs, glassware, sugar, and exotic nuts, to name a few. In this way, many items that could not be produced in Northern Europe were introduced to those who could afford such luxuries. Medieval fairs were not only opportunities for local goods to be exchanged, but also a chance to see and buy exotic goods from distant lands, In the age before the supermarket and the mall, the trade fair was at the cutting edge of commerce.

In 1066, the Normans conquered England. The Norman French established a new system of trade fairs across Britain that lasted into the modern era. The medieval kings of England granted charters, or the right to hold a trade fair, to certain organizations. One of the earliest charters was given to the Priory of St. Bartholomew in London in 1102. In many cases the charters were given to church organizations to curry favour with this important institution. It was an advantage to have the church on your side in medieval England.Other charters were granted to nobles to reward loyalty or gain influence. But the charters also had a more practical side. These trade fairs generated income, not only for the church, but also for the king. Tolls (admissions) and stall rentals generated cash for both groups. St. Ives Fair paid the king a flat fee of £50 per year. It was also the chief source of revenue for the monks of nearby Ramsay Abbey, and a staff of twenty monks spent the entire year planning and running the fair. The king was careful not to grant too many charters. An annual cycle or circuit of fairs was set up to allow merchants to attend the more important shows. The larger fairs were Stamford (Lent), St. Ives (Easter), Boston (July), Winchester (September), Northampton (November), and Bury St. Edmonds (December). Other smaller fairs filled in the dates between the large fairs. Most of the smaller fairs were held at the manor houses of the nobility who were granted the charter. Records of these ancient charters survive to the present day. A good example is the charter granted to Lymington Fair in the County of Hampshire, which reads as follows:

"Charter for St Matthew's Day (21 Sept); granted 12 June by King Henry III to Baldwin de Insula, son Baldwin de Insula, sometimes Earl of Devon. To be held at the manor."[1]

Later records show the fair at Lymington was a going concern in the late thirteenth century. The twentieth-century fair, held the first week of October each year, appears to be a continuation of this medieval fair. They were primarily cloth fairs, established for the marketing of sundry fabrics. But countless other goods ranging from horses, firewood and grain, to exotic wares from across Europe, were also bought and sold.

Political and economic changes had a strong effect on these cloth fairs. As the economy of England changed, the foreign cloth merchants began to be replaced by English merchants, and the variety of goods expanded. The English merchant class now dominated the fairs, and the growing towns (or boroughs, as they were known in the Middle Ages) began to dominate the fair circuit. In the 1530s, the monasteries and abbeys were dissolved in the English Reformation. Charters to hold trade fairs were transferred to the boroughs and nobles. Often the local merchants became responsible for the fairs. The religious element was removed, and the entertainment element began to creep in.

Countless acts and laws were passed to regulate and control fairs in the kingdom of England. The holders of the charter were given the power to set up courts of justice during the fair. These were called the courts of Pie Powder, after the French term *pied poudreaux*, meaning dusty feet, because peddlers, merchants and travelling salesmen could be identified by the dust on their boots. The courts were presided over by local magistrates or stewards, and since the fairs lasted at best only two to three weeks, punishment needed to be fast and effective. Fines, public whipping or a session in the stocks were common. In keeping with the carnival atmosphere, punishment was sometimes geared to fit the crime. Brewers were made to drink their own bad beer or bakers to eat their mouldy bread.[2] A hangman caught stealing at St. Bartholomew's fair was hanged.[3] But most of the court cases involved petty squabbles over prices, money and payment of debts.

Travel to and from the fairs was a risky business. The holders of the charters were required by law to provide a set number of constables to guard the roads and keep highwaymen and common thieves from preying upon both fair-goers and merchants. Likewise, the fairgrounds were often infested by pickpockets, cutpurses and shoplifters, so even more constables patrolled the grounds to discourage these nefarious criminals. At five-minute intervals these policemen would bellow, "Watch out about you!" to startle both predators and prey. At night, a large flotilla of nightwatchmen would patrol the grounds, safeguarding the merchants' stalls. The court of Pie Powder swiftly administered justice to those caught in the act. Imprisonment, cutting off an ear or a finger, and even execution, were not uncommon.

As the trade-fair system became established in England, specialty fairs began to grow. The earliest fairs were cloth fairs, but other commodities also had their day. Horse fairs and livestock fairs are two that have continued to the present day. The oldest continually held fair is said to be the Lee Horse Fair in Yorkshire, which started in 1138. A goose fair at Nottingham featured as many as 20,000 geese (for sale). Another town held a horn fair, which featured items made from animal horn. Mop fairs were once held in rural England. These various fairs were among the earliest employment agencies. Agricultural workers advertised their service at these hiring shows. To advertise their professional specialties, workers carried a badge of their trade; for example, a shepherd tied wool to his hat, a carter carried a whip, a thatcher pinned straw to his coat, and a housemaid held a mop or broom.[4] The badges of some professions will be left to the reader's imagination! Many unhired labourers trudged from fair to fair looking for employment.

The two most famous trade fairs in English history were St. Bartholomew's Fair at London (1102–1855) and Sturbridge Fair near Cambridge (1211–1855). These two were noted for their longevity, and they maintain a mythical status among fair historians. They also bridged the gap between medieval trade fairs and modern agricultural and commercial fairs.

St. Bartholomew's Fair

The Fair of St. Bartholomew was held on St. Bartholomew's Day, August 24, and was sponsored by the priory of the same name. Originally a trade or cloth fair, it quickly evolved into much more. Situated in the well-populated capital city of London, the fair was used for many purposes. Slaves were sold and heretics burned at the stake. The king of England used the draw of the fair to make political statements to the crowds. In 1305, the Scottish patriot (or traitor, depending upon your point of view) William Wallace was executed at St. Bartholomew's Fair for the "entertainment" of the fair-goers. The king maintained a gallows at St. Bartholomew's and used it frequently as a reminder to his subjects of the importance of law and order, and to show who was the boss. Usually held yearly, the fair was sometimes cancelled due to external influences such as civil war or disease. The Black Death, or bubonic plague, caused a disruption in the fair cycle. The 1348 fair was dampened by the fact that 50,000 burials had been made in the past year in a graveyard bordering the fairgrounds. As one chronicler put it, "What mirth was there in that handful of the living camped so near the silent congregation of the dead?"[5]

An English fun fair of the 1700s. Music and mayhem and not a lot of agriculture or education. In the age before the camera, history was recorded in paintings. This work was executed by the famous English artist Hogarth.

Another important element of the Fair of St. Bartholomew was the presence of the theatre. Being in London, the fair was frequented by troupes of actors. The first performances were recorded in the 1300s. The early performances were mainly miracle or passion plays with religious themes: a good idea, considering the proximity of both the priory and the gallows! An early group was the Company of Parish Clerks, an incorporated company or guild that had charge of the records of the burials and the births in London. They performed "Matter from the Creation of the World" at a fair booth. (Evidently the job paid rather poorly, and further income was necessary). With the dissolution of the Priory in 1538, restrictions on the stage shows were relaxed and an anything-goes atmosphere prevailed. Because London was the capital of the kingdom, political satires became the norm. This led to the amusement of the masses, but resulted in the resentment of certain groups, especially those being satirized. Consequently a series of regulations was passed aimed at controlling "dramas and plays" at the fair and urging the fair to return to its mercantile roots. But St. Bartholomew's Fair was too important an outdoor theatre, and the entertainment remained.

St. Bartholomew's Fair also attracted a less cultural dimension of the entertainment world. Jugglers, clowns and other entertainers had always been a part of the fair, but in the 1700s freak shows, exotic-animal displays and other sideshows became common. Games of chance appeared, followed by the forerunners of today's midway rides. The trade aspect of the fair declined and was replaced by carnival. The main features of the fair became performances by actors, acrobats and menageries. The main acting company at the time was Richardson's Traveling Fairground Theatre. The two largest menageries were Wombwell's and Atkins. Sadly, the conditions for the exotic animals, such as zebras, pelicans, lions, tigers and hyenas, were dismal. Wombwell's elephant died while walking to London one year. To profit from this misfortune, the owner set up a separate sideshow featuring the dead elephant. It was reasoned that a dead elephant was rarer than a live one! I trust, or rather hope, that this show was not repeated at other fairs.

A list of cash returns for the 1827 edition of St. Bartholomew's fairs tells the story of the changing role of this institution.

Wombwell's Menagerie £1,700

Richardson's Theater £1,200

Atkins Menagerie £1,000

Morgan's Menagerie £150

Exhibition of "pig-faced lady" £150

fat boy and girl £140

head of Wm. Corder, Quaker, hanged for the murder of Marie Martin—the crime being revealed through a dream of the victim's mother £100

Ballard's Menagerie £90

Ball's theater £80

diorama of the Battle of Navarino £60

Chinese jugglers £50

Pike's theater £40

a fire-eater £30

Frazer's theater £26

Keyes & Line's theater £20

a Scotch giant £20 /[6]

Long gone was the market fair, replaced by a carnival of the odd and unusual.

Sturbridge Fair

Sturbridge Fair was a rural trade fair, held in a farmer's field near Cambridge. The original charter was granted by King John in 1205 to the leper hospital of St. Mary Magdalen, with the proceeds from the fair to be used to support said hospital and the Priory of Barnwell. By 1496, Stourbrige Fair, as it was then known, was leased to the burgesses or merchants of nearby Cambridge, who, upon the dissolution of the Priory in the 1530s, acquired full rights to the fair. From earliest times, the fair attracted a variety of merchants from all over England and overseas. Its main commodities were agricultural, the big three being wool, hops and horses. But even rural

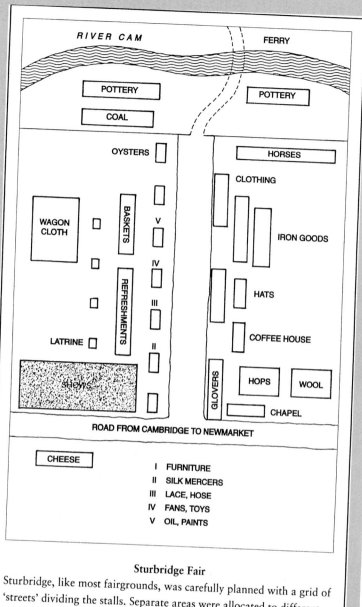

Sturbridge Fair
Sturbridge, like most fairgrounds, was carefully planned with a grid of 'streets' dividing the stalls. Separate areas were allocated to different goods and trades. Since few people could read, large signs were hung over the stalls. A boot stood for a cobbler, a knife for a cutler, a key for a locksmith. At the centre of the ground was a huge square called the **Duddery** where clothes and fabrics of all sorts were sold.

Sturbridge Fair was not without its problems. Concerns that the carnival/sideshow element was subverting the trade show were expressed as early as 1604, when King James I presented a mandate to the mayor of the borough of Cambridge ordering that they, "refrain, inhibit and forbid as well all manner of unprofitable or idle games or exercises ... whereby throngs concourse or multitudes are drawn together or whereby the younger set are, or may be, drawn or provoked to vain expense, loss of time or corruption of manners, as also to publish, act, set out or make any such unprofitable or idle games."[7]

However, the invocation was more easily written than enforced, for a 1693 poem contains the following lines:

"A fire licking a child's hair was to be seen at Sturbridge Fair with lambent flame, all over a sweating mare. Women dancers, puppet players, at Bartholomew and Sturbridge Fair."[8]

Some problems seem to be timeless

By the 1700s, Sturbridge Fair, with or without its carnival midway, was considered a world-class fair. The famous writer Daniel Defoe left this report of his visit in 1723:

"The last day of the Fair is the horse fair where the whole is closed with horse races, to divert the meaner sort of people only, for nothing considerable is offered of the kind.

"This ends the whole Fair and in less than a week there is scarcely any sign left that there has been such a thing there."[9]

Nottingham Goose Fair

Not all English fairs disappeared over time. The Nottingham Goose Fair, originally chartered in 1284, is an example of an ancient fair still operating today. The Danes established a market fair at Nottingham over one thousand years ago, but the current fair dates from a charter granted by King Edward I in 1284. The fair was held on St Matthew's Day (September 21). Tradition maintained it was good luck to eat goose on this holy day. Consequently, thousands of geese were offered for sale at the fair. Many "waddled" long distances to the fair, their feet coated with sand and tar to reduce wear and tear. With the Protestant Reformation of the 1500s, certain religious holy days fell into disfavour. Goose sales declined and were replaced with other commodities such as cheese. In 1764, a 33 per cent increase in the price of cheese led to riots. When an angry mob looted the fairgrounds, the mayor of Nottingham stepped in to try and quell the rioters. A huge cheese wheel was rolled over the mayor, but his injuries were more to his pride than his body. Gradually, the market fair evolved into a fun fair. Madam Tussaud's wax museum appeared at the Nottingham Goose Fair as early as 1819. Wombell's Menagerie was a regular sideshow as early as 1805. The first roundabout, an early hand-turned carousel, appeared in 1855. The first Ferris wheel (from America) was presented in 1906. The Nottingham Goose Fair has been held every year since its charter, with only eleven exceptions. World Wars I and II led to cancellations, as did periodic outbreaks of the bubonic plague. When Nottingham City Council tried to move the fair from its traditional site, a protest rally led by the president of the Showmen's Guild drew 12,000 angry citizens. The presence of many huge riding devices had overflowed the old market square, making a move necessary. The present-day show features 55 adult rides, 40 children's rides, 225 games of chance and approximately 400 vendors peddling a wide variety of goods. In 1994, the Nottingham Goose Fair (minus the long-gone geese) celebrated its 700th birthday.[10]

Demise of the Great English Fairs

The 1800s saw the decline and demise of the great English fairs. Many had changed from trade fairs to fun fairs, a term commonly used in Victorian England. The market element of the fair had been usurped by the growth of retail stores. The retail storekeeper was open for business all year round and carried a variety of goods that was unparalleled. Also, amusement parks, zoos, circuses and theatres had now developed more permanent sites. The agricultural fairs were still held in rural areas, but permanent markets and sale barns were also damaging them. Britain was becoming an urban nation, and the traditional English trade fair was doomed by this shift. The agricultural fairs continued in the diminished rural areas. In 1855, both Sturbridge and St. Bartholomew Fairs were held for the last time. An era had ended, but another era was born.

An early English dobby or roundabout, circa 1840. The earliest ancestors of the merry-go-round or carousel were hand driven. Volunteers gladly turned the crank in exchange for free rides. Horses could be utilized, but why bother, when free labour was so plentiful. Steam engines replaced muscle power in the 1860s. The increased horsepower meant the riding devices could be made bigger and more elaborate. By the 1890s, the merry-go-round as we know it was standard operating fare at any respectable carnival midway.

World's Fairs

Also affecting trade fairs was the growth of the giant industrial exhibitions. The concept of a world's fair had started in 1798 in Paris. The first exposition lasted three days and had 110 exhibitors. In 1849, a similar show ran two months and had 4,494 exhibitors. In 1851, London sponsored the great Crystal Palace Exhibition. This industrial show was meant to showcase the best products of British industry — which sounds a little like the goals of medieval fairs. The great Crystal Palace hall was 1,604 feet long by 384 feet wide and cost £193,000 to build. It covered 21 acres and held 13,937 exhibitors. This world's fair lasted five months and attracted over 6 million fair-goers! A new standard had been set in the world of trade fairs.[10]

Other nations rushed to match the standard set by the Crystal Palace Exhibition. Countries around the world took turns hosting world's fairs after 1855. Huge trade fairs or expositions were set up all over Europe and, indeed, all over the world. Travel became more practical with the invention of the steamship and the railway, and exhibitors and fair-goers from around the world could now travel easily and efficiently from continent to continent. The world of trade and technology was changing so fast that many people needed to attend these fairs regularly just to keep up with the new advances. The spread of literacy and the printed word created societies keenly interested in new technology and new products from around the world. The world's fairs became educational as well as commercial fairs. It became a source of honour and pride for a nation to stage a world's fair. Each nation's exposition attempted to outdo the previous shows.

The Agricultural Revolution

The agricultural sector had always been a big part of medieval fairs. In the 1700s an agricultural revolution swept through Britain. Agricultural improvement and scientific farming became fashionable among progressive-minded individuals in the British land-owning classes. Better machinery, better methods of farming and improved livestock breeding increased the wealth and prestige of progressive farmers. Thomas Coke was an English landowner who sponsored conferences dedicated to the exchange of ideas on agricultural improvement. Royal support came from King George III, who was nicknamed Farmer George. He established a model farm at Windsor Castle, where he implemented many of the new ideas. The first English Board of Agriculture was formed in 1793. The Royal Agricultural Society held its first fair in 1839. Agricultural improvement societies sprang up all over the British Isles. They had many ways of promoting agricultural improvement. They held seminars, published books and journals, and sponsored model farms. But their greatest successes were achieved at the agricultural fairs. Building on the rich tradition of trade fairs, the agricultural revolution was reflected in the fairground.

Livestock competitions grew to be a main segment of agricultural fairs. Instead of just housing meat markets, the fair became a venue for showing off better stock and encouraging the breeding industry. On record is the Tully Hereford steer that stood an unprecedented 6 feet, 7 inches high and captured first prize at the Smithfield Fat Stock Show in 1799. The famous "White Heifer that Traveled" won countless prizes in

Hand-drawn poster promoting an early (pre-1840) Canadian fair. The motto "God Speed the Plough" is an old English prayer used to bless the farmers at Michaelmas church services.

show rings across England. After victory in the show ring, the proud owner would fill a depression in the cow's back with rum, and the competitors would celebrate by drinking the rum (and a few cow hairs) by straw. The English concepts of honour and pride certainly transferred to the show ring. When the famous Thomas Bates was challenged by another shorthorn breeder at the English Royal Stock Show in York in 1842, Bates walked his prize cow, Old Brokenleg, forty miles to win the challenge and keep his honour intact.[11]

The Move to North America

The British settlers who came to North America after 1700 brought two great agricultural traditions: the market fair and the agricultural improvement society. In pioneer North America, the farm was king; agriculture dominated. Early trade or cloth fairs and carnivals were almost unknown. Farming was everything to the new settlers. Market fairs operating in North America before 1850 were almost exclusively agricultural market fairs. The first agricultural market fair in North America was held at Windsor, Nova Scotia, in 1765. Agricultural societies were soon established in the United States at Charleston, South Carolina, and Philadelphia, in 1785. New York and Boston soon followed in 1791 and 1792, respectively. Like their early Canadian counterparts, these agricultural societies were formed by gentleman farmers and merchant groups. They advanced the cause of scientific agriculture and the business of agriculture rather than concentrating on practical advances in farming. These early societies were largely failures.

The father of American agricultural societies was Elkanah Watson. A New England wool merchant, Watson's mills were constantly plagued by a shortage of good-quality wool. To remedy this problem, he began to encourage improvement in breeding sheep. Watson tested public support by tethering several of his prize Merino sheep in the town commons for a day. All interested parties were urged to attend and inspect this new breed of sheep imported from Spain, which was highly prized for the quality of its wool. The next year, other breeders of Merinos were invited to display their animals as well. This led to the development of an early agricultural fair, dubbed the Berkshire system of fairs after the county of origin in Connecticut. The first Berkshire fair was held in 1811. Approximately $70 in prize money was offered and 3,000 farmers attended. Watson wanted to encourage "practical" or "real" farmers to belong to his societies and be instructed in agricultural improvement. The Berkshire system reached its zenith in the 1820s and was especially dominant in New York State, just next door to Upper and Lower Canada.

Each Berkshire fair followed a set format. The one-day shows opened with a parade of livestock to the fairgrounds, usually the town commons.

Very early (pre-1900) Ferris wheel. This hand-cranked model only held four seats and is very primitive by today's standards. But when daddy wanted to take a photograph of his two sharply dressed sons, this marvel of the times was selected for the backdrop. Guess which part of the fair impressed this family.

The animals were immediately judged. Spectators lined the show ring to watch the judging. Spectator comments were loud and often critical. Grain and other articles were judged next, again in full public view. The Berkshire system encouraged all manner of competitions, including "ladies' work" and domestic sciences. Next on the agenda was a plowing match. It was followed in the late afternoon by livestock sales. Then all interested parties adjourned to a local tavern for a banquet. After-dinner speakers then presented lectures on sundry agricultural topics. A dance or ball closed the fair day.

The Berkshire agricultural societies became so influential that state governments extended substantial grants to their fairs. The State of New York granted $10,000 to Berkshire fairs in their peak year, 1819. While very influential among farmer's groups, the Berkshire fairs suffered from poor finances. Even with government assistance, these purely agricultural shows were money losers. In 1825, government assistance was cut off and the Berkshire fairs began to disappear. These exhibitions were totally dependent upon government funds to pay their prize lists. Left to their own devices, they went bankrupt. But the demise of the Berkshire fairs was not the end of agricultural societies in the United States. The Berkshire system simply set the stage for the state/county system of agricultural fairs that grew up across the United States in the next half century. Although there were never any Berkshire societies in Upper Canada, they did have an influence in that colony. Early agricultural societies adopted many of their ideas. The Canadian variations did not use the parade or the evening ball. But the livestock shows, the evening banquet and the lectures were very common among Canadian agricultural societies prior to 1850.

The collapse of the Berkshire system did not stop the promotion of agriculture. American government authorities began to realize the value of agricultural societies. In 1841, the New York State Legislature granted

$8,000 for the promotion of agriculture in the state through an annual fair. The first New York State Fair was held at Syracuse, a major transportation junction in northern New York State. Approximately 5,000 farmers attended. In a pattern repeated across Canada, the state fair led a transient existence for the next forty-nine years. It moved from town to town, visiting a different site every year. Likewise, the provincial exhibitions across the border in Upper Canada also travelled from site to site. In 1890, convinced of the value of having a permanent fairgrounds with better facilities and buildings, Syracuse was selected as the permanent site for the New York State Fair. It has remained there to this day.

Following the New York pattern, a system of state/county fairs spread across the United States. In a pattern similar to that in Canadian history, settlers from the eastern United States transplanted their love of fairs to newly settled regions in the West. Each state sponsored a large state fair, where the winners from the various county fairs competed. By the 1870s, the United States had over 1,300 agricultural fairs. This number peaked in the 1920s, when over 3,000 fairs were held every year across the United States. Many historians consider the era from 1850 to 1890 to be the golden age of American agricultural fairs. Shows held during this era were purely agricultural and education-oriented. Countless new inventions were first demonstrated at these fairs. For example, the threshing machine, a great advance in labour-saving for the farmer, was first displayed at a fair in 1878. The threshing machine had actually been patented ninety years before, but was virtually unknown until Jerome Case began to tour the fair circuit with his model.

So, like their Canadian cousins, American fairs started off as agricultural showcases with a smaller commercial/entertainment sector. However, as the twentieth century advanced, the entertainment sector became more and more important, often at the expense of the agricultural groups. The decline in the number of farmers and farms, as well as a growth in urban centres, meant American fairs had to adjust their focus or change to address these developing trends. Agriculture is still a big factor in most American fairs, particularly county fairs. However, urbanization across the country led American fairs to change their focus to try to attract the urban market. American fairs enjoy a sacred place in American culture. References to fairs are quite common in music, movies, books and many other vehicles of popular culture. They are ingrained in American culture, just as fairs are in Canada. In 1986, 2,300 American fairs attracted a total attendance of approximately 125 million fair-goers. In many ways, the history of Canadian and American fairs are strikingly similar. They face many of the same problems and issues. And yet again, they differ, because as we have said, fairs reflect the communities they serve.

COME TO THE FAIR
September 20th to 30th, 2001

- Thrill to rodeo, show jumping and exciting horse events in the main arena
- See the spectacle of champion cattle, horses, alpacas, etc.
- Watch your favorite dog strut its stuff in the biggest dog show ever
- Pat an alpaca or a lamb or a pig in the famous animal nursery
- Cuddle a guinea pig or rabbit in the Pet Precinct
- Watch a cow being milked and follow the milk trail in Mootown
- Marvel at the orchids and the best fruit and veggies in Horticulture World
- See the chips fly in the spectacular wood chopping competitions
- Search for a bargain in the commercial pavilions
- Gasp at the stars in the X S zone with Extreme Riders on the vertical ramps
- Tap your toes to the best music in the Country Music Muster
- Thrill to the rides, choose a showbag and finish the day with fireworks

Guess which fair this 2001 ad is for. It could be any fair in Canada or the US. But it is actually an advertisement for the Royal Melbourne Fair in Australia! Any fair-goer in North America would feel totally familiar with such a lineup. Agricultural fairs, as we know them, are common in many parts of the world.

The British tradition of agricultural societies and fairs has spread all over the world.

British settlers carried these ideas to far-flung corners of the globe. A study of the history of the Melbourne Agricultural Society in Australia will reveal a surprising number of similarities between Canada and Australia. By 1840, a steadily growing population in the Victoria District led to many progressive farmers demanding that an agricultural improvement society be formed. There was a perceived need for better farming practices and more education for farmers. Drawing upon their British traditions, a group of interested settlers established the Pastoral and Agricultural Society of Australia Felix in 1840. The first fair in 1842 was a failure, but the directors persisted, and after 1848 the Melbourne Fair became an annual event. There are many similarities between the history of the Melbourne Fair and any Canadian fair. Here are a few historical milestones for the Melbourne Fair:

1848 first plowing match

1855 government land grant to buy fairgrounds

1874 sheep-shearing demonstration

1890 Queen Victoria gives permission for prefix "Royal" to be used

1904 arts and crafts pavilion constructed

1910 first parade

1911 first women's division

1915 grounds requisitioned by army during World War I

1931 salaries cut due to Great Depression

1937 first nighttime show

1938 Showmen's Guild formed to regulate midways

1939 World War II, army uses grounds until 1946

1951 Olympic jumping and dressage events held

1957 first Fair Queen competition

1964 animal nursery

1968 first Sunday show

1971 last Midway "tent show" (Jimmie Sherman Boxing show)

1981 first timber show

1986 Australian beer awards

1992 dairy produce competition

1998 150th anniversary

Can you see similarities to the history of Canadian fairs? It's a small world after all![1]

PLOWING MATCHES

One early activity of agricultural societies was organizing plowing matches. As was the case with fairs, this idea came from Britain. In England, plowing matches were traditionally held on the first Monday after the Epiphany, which was considered the first day of the farmer's year. This ancient prayer led off the festival:

"In fair weather and in foul, in success and disappointment, in rain and wind or in frost and in sunshine; Godspeed the plow."[1]

"Godspeed the plow" was the motto adopted by several early agricultural societies in Eastern Canada. Proper cultivation technique was a noble cause adopted by many early agricultural societies. In 1819, eight plowing matches were held in Nova Scotia, and one in Quebec. In 1824, the Northumberland Agricultural Society in Ontario held the first match in that colony. Other provinces followed as settlement advanced: Prince Edward Island in 1837, British Columbia in 1870, Manitoba in 1874 and Alberta and Saskatchewan in the 1890s, all began to sponsor plowing matches.

By 1871, fifty Ontario agricultural societies were sponsoring plowing matches, usually held separately from the fair. It was too demanding on time schedules to hold both plowing competitions and fairs on the same day and at the same site. In the age before tractors, horses and oxen were the beasts of burden commonly used. An 1840 plowing match contained the following classes:

Class 1 – open

Class 2 – open to all, except to those of European extraction

Class 3 – lads, 17 to 29

Class 4 – lads, under 17

Prizes: 1st – 1 pound; 2nd – 15 shillings; 3rd – 10 shillings

Doukhobor women pulling a plow, Saskatchewan, early 1900s. There has never been any class at a plowing match for fourteen-woman hitch. These Doukhobors are plowing unbroken land in Saskatchewan in the early 1900s. Horses were either scarce or too expensive, so the largest pool of untapped labour was utilized: the women of the Doukhobor colony. Or maybe horses were just too valuable to risk?

A typical contest in 1866 had the following rules:

1. each plowman assigned 1/4 acre
2. each allowed one assistant to set marker stakes
3. start 11:15 a.m., finish by 3:15 p.m.

(Four competitors did not finish!)

Prizes awarded included:

Fanning mill
Horse cutter
Harrows
Cook stove
Saddle
Suit of clothes
Purebred ram
Carton of sheep-tick destroyer
Stable lantern
Silver-mounted harness
Table lamp
Scotch collars
Melodeon
A pump
Buggy
Straw cutter
Chest of drawers
Pair of boots

A special allowance was made for anyone using oxen! Obviously they needed a little more time to complete the course.

In the first decade of the twentieth century, tractors made an appearance. Winnipeg fair sponsored a tractor contest in 1908 to allow manufacturers of the new-fangled devices a chance to show off their products. Nine tractor makers participated. The machines, all steam-driven, were judged on:

Weight
Horsepower

Fuel and water consumption
Distance on a tank (of water)
Turning range
Protection of parts from mud and dust
Accessibility of parts
Speed
Ease of manipulation
Clearance from ground
Steadiness
Price FOB Winnipeg

Each entry had to plow, disc, haul loaded wagons and perform belt-work. Three tractors broke down during their demonstrations, but two were repaired on the spot. Unfortunately, heavy rains caused the tractors to flounder "like elephants in a swamp." First place was captured by a Kinnear-Haines 30 horsepower, second was a 15-horsepower International Harvester, while third place went to a 30-horsepower Marshall. The contest was so successful, it was repeated at the next six fairs. The 1913 show attracted 25 entries, but only two were steam driven: times were changing![2]

In a bid to improve the organization of plowing matches in Ontario, the Ontario Ploughman's Association was set up in 1911. Its main goal was to encourage county plowing matches and organize an all-Ontario meet, later to become the International Ploughing Match and Farm Machinery Show. The first match was held on the future site of Sunnybrook Hospital in Toronto in 1913. The contest attracted thirty-one single-furrow horse-drawn entries. There was no class for tractors yet! In the 1915 event, eight tractors participated while twenty-five plowed in 1919 and fifty in 1920. At the 1930 match held at Stratford, Ontario, attendance exceeded 100,000 — a real milestone. There were 307 horse-drawn entries and 132 paticipants in the tractor classes. The International Ploughing Match and Farm Machinery Show has gone on to become a major agricultural event, closely resembling its cousin, the agricultural fair.

The Ploughman, Brigden, Ontario, Fair. Titled *Early in the Morning, Down on the Farm*, this 18-foot-high metal sculpture was designed to show the relationship between plowing, farming and fairs. It stands at the front entrance of the Brigden Fairgrounds as a reminder to all who enter of the bond between agriculture and fairs.

Canada's Fair History

THE HONOUR OF HOLDING the first agricultural fair in Canada belongs to the Hants County Agricultural Society of Windsor, Nova Scotia. On May 21, 1765, local farmers gathered on Fort Edward Hill for a fair. It was held in the spring, to enable farmers to buy and sell livestock just before the breeding season. The prize list did not specify livestock breeds, just the age and sex of the animal. Clearly, the organizers were happy to just have farm animals of any breed at this first fair. Prizes consisted primarily of donated items, which reduced the strain on the society's finances, but also led to some interesting matches on the prize list. Another curiosity was the presence of entertainments such as horse racing, wrestling and marksmanship. This earliest of Canadian fairs was clearly meant to be a community fun day as well as an agricultural-improvement exercise.

The fair at Windsor announced itself with the following proclamation:

"Whereas it is thought the establishing of the fair at Windsor will be of great utility to the province of Nova Scotia, a number of Gentlemen of Halifax being desirous of promoting every measure that may induce to the public good, have entered into a subscription for premiums and rewards and will cause the following to be given on Tuesday, the 21st of May, 1765, the first day of the fair.

"To the person who shall bring to the fair, Prize

For sale of the greatest number of Neat cattle	3 yds of English broadcloth & a silver medal
Greatest number of horses	a saddle, bridle and a medal
The greatest number of sheep	a pair of shears, a pair of cards, medal
The largest pair of working oxen	a plow-share [sic] and a medal
Next largest pair of oxen	a plow share [sic]
The finest and largest cow	a butter churn and a medal
The largest yearling bull	a medal
The largest yearling heifer	a medal
Largest pair of steers	a medal

Best pair of harnessed horses	a plowshare and a medal
Next best pair of harnessed horses	a plowshare
The best single horse	a medal
Next best single horse	a whip and a pair of spurs
Largest and fastest wether sheep	a pair of wool cards and a medal
Best butter not less than 12 lbs.	6 yards ribbon and a medal
Best cheese not less than 12 lbs.	6 yards ribbon and a medal

The day of the fair will be given for the following diversions

The best running horse	a pair of buckskin breeches and a medal
Next best running horse	a pair of buckskin breeches
Next best running horse	a whip and a pair of spurs
Best wrestler under 25 years old	a laced hat and a medal
Next best wrestler	a pair of shoes and buckles
Next best wrestler	a pair of buckskin gloves
Person that shoots at a mark with a single ball the best in 2 shots at 80 yards distance a neat and a medal	a neat and a medal

"Proper persons will be appointed to judge the different rewards and to regulate the diversions.

"The medals will be of silver the size of a dollar with an inscription of thereon suitable to the occasion.

"Three or four of the best stallions in the province will be provided gratis. Many of the subscribers and others will be at the fair to purchase a great number of cattle."

Halifax, the 29th of March, 1765[1]

Agricultural Societies in Atlantic Canada

To push the cause of agricultural improvement, the various colonial governments established central or provincial agricultural boards or societies and endowed them with government funds to promote agriculture. These central boards imported breeding stock, published educational materials, held livestock shows and supervised the local/county agricultural societies. But it was the local societies that organized the agricultural fairs. The central or provincial boards came and went, but the local societies kept going through good times and bad. Nova Scotia's first provincial society was established in 1789, New Brunswick had a provincial board of agriculture by 1790, and Prince Edward Island by 1827. These provincial societies were hamstrung by many difficulties. Transportation was difficult in pioneer times, and the central boards tended to concentrate around the capital city at the expense of the outlying districts. The members were mainly businessmen and merchants, gentleman farmers rather than practical farmers. And finally, these boards were totally dependent on government financing. When grants fluctuated, so did the provincial boards.

Royal North West Mounted Police musical ride, CNE, 1920. The RCMP Musical Ride is a very old institution. Its origin lies way back in the days of the old NWMP. Notice the officers' white colonial-style pith helmets.

Local societies in Atlantic Canada continued to grow in numbers. By 1840 there were thirty societies in Nova Scotia and eleven in Prince Edward Island. These groups had started plowing matches in the year 1818 in Nova Scotia, and field-crop competitions in 1825 in New Brunswick. The importation of breeding livestock was declining as private businesses replaced agricultural societies in this field. Similarly, farm implements and seed grains were now the domain of the private entrepreneur. This was not a bad trend for agricultural societies. Importing livestock was a risky, money-losing business. Stud fees and funds acquired from auctioning off these animals, did not cover the costs of keeping them, let alone the return of the original investment. The New Brunswick Central Agricultural Society lost a £1,500 investment in Clydesdale stallions when the ship carrying them sank en route to Saint John. Combined with sporadic government funding, these reasons led the Atlantic agricultural societies to abandon the livestock-importing business and concentrate instead on holding agricultural fairs. This pattern was to be repeated over and over again across Canada. Newly created agricultural societies originally participated in livestock breeding but eventually fell back on fairs as their main activity.

Newfoundland, Britain's oldest North American colony, was also the least agricultural. The earliest settlements, almost exclusively fishing stations, imported most of their food. By the 1840s, a new governor, Sir John Harvey, encouraged the formation of an agricultural-improvement society. He enthused, "Newfoundland is, in reality, something more than mere fishing stations, and possesses resources beyond the mere rocks on which to dry the nets of the fisherman." With this speech, a Newfoundland agricultural society was formed in St. John's on January 14, 1842. At the time, a few farmers in the Avalon Peninsula were the main supporters of such a society. This society promptly imported some Ayrshire cattle from Prince Edward Island and Nova Scotia to kick-start a small dairy industry. Grain production was also encouraged, to make the colony more self-sufficient. A list of competitive classes from the 1848 field-crop competition reads:

1. The Lemarchand Cup for the greatest breadth of wheat crops, fair marketable quality for two consecutive years
2. for the best crop of wheat, on land of not less extent than 3 acres £6
3. for the best crop of wheat on land of not less than 2 acres £4
4. for the best cultivated crop of wheat on land of not less than 1 acre £3
5. for the best cultivated crop of wheat on any land of not less than an acre £3
6. for the best crop of oats on any land of not less extent than 2 acres 3 £[2]

Parade of champions, Fredericton Fair, New Brunswick. The livestock are lined up by breed, with mature bulls leading the way. Holstein cattle had made an appearance in Canada by this date.

Note the small acreages involved. The number of agricultural societies was hampered by the small number of actual farmers. The more remote outports of the colony frequently complained that government funds were being spent exclusively in St. John's. An act was passed in 1856 allowing for the sum of £250 to support the agricultural society activities. It came with the following strings attached:

"The sum of £150 in the purchase of seed and of cattle, to improve the breed, in and for such Outport Electoral Districts as may require the same; and the remaining sum of £100 to be expended for like purpose in St. John's."[3]

The amount of suitable land for agriculture was very limited in Newfoundland. Farming was never a big industry. Consequently agricultural societies and fairs were never very numerous. To date, there are five fairs and five festivals in the province.

Agricultural Societies in Quebec

British settlers brought agricultural societies to Quebec after 1783, with the hope of encouraging better production from existing lands, as well as opening up new areas. The Agricultural Society of Quebec was established in 1789, followed closely by the Montreal Agricultural Society.

Stage and the judges' stand, Sherbrooke Fair, Quebec, 1912. Pictured is a well-organized stage with hanging backdrops resplendent with scenery, and a complete stage band in the pit in front of the stage. Clearly, some sort of pageant or stage show is being set up. Numerous livestock and horse rings are in the background, along with the livestock barns. The grounds are a good example of a well-organized, well-run fairgrounds. Oh yes, and the crowd is intent on something coming up the track: a horse race perhaps?

The spread of local societies was slow due to distrust between French and English factions, but the new ideas led to a gradual spread of agricultural societies throughout Quebec. In 1818, the legislature granted £800 to Quebec, £400 to Trois-Rivières, and £800 to the Montreal agricultural societies. By 1849, there were only sixteen societies in Canada East, as Quebec was called at that time. As was the case in Atlantic Canada, the earliest societies dabbled in more than just fairs. In 1847, the central board or society was given the following operations budget by the government:

£4,000 to agricultural societies for prizes at fairs
£1,500 to agricultural schools and model farms
£600 for essay contests
£700 to publish a farm journal
£700 to hire two superintendents of agriculture

While fairs were clearly the biggest operations, the agricultural societies were also responsible for farm journals, model farms, agricultural schools, agricultural education and a staff to oversee these functions. However, times and attitudes quickly changed, and by 1851 all the funds granted to Quebec agricultural societies were invested in fairs and prize moneys. These fairs were originally highly specialized. In 1831, there were eight county livestock shows, one general cattle show, one fat-hog show, and one fat-ox show, as well as general fairs held across Lower Canada. The Montreal Fair of 1831 offered prizes in 220 classes.[3]

The history of the Lachute Fair is a good example of growth and change in Quebec agricultural societies, and indeed in fairs across Canada. Founded in 1825 to "improve agriculture," the society endured many name changes. The original name "County of York Agricultural Society" was changed to the "Two Mountains Agricultural Society" and finally to the "County of Argenteuil Agricultural Society." The earliest efforts were geared to spring cattle shows and fall plowing matches. Fairs alternated between the centres of St. Andrews and Lachute. By 1883, the directors decided to forgo the transient nature of their fair and set up in a permanent place. Three offers were received from the villages of Lachute, St. Andrew's and Chatham. Lachute was chosen for two reasons: it was a central site for the whole county, and it was served by a railway. Before 1879 there was no admission charge to the fair. By this date, expenses had increased to the point at which the directors decided that a 10¢ admission fee was necessary to keep the society solvent. When the federal government cancelled grants due to wartime situations, the 1940 fair was discontinued. Realizing their mistake in cancelling the fair, the directors reconstituted the fair the next year. In 1964, a chunk of the fairgrounds was sold to the local Lions Club for erecting an arena. The history of Lachute Fair is typical of fairs across Canada.[5]

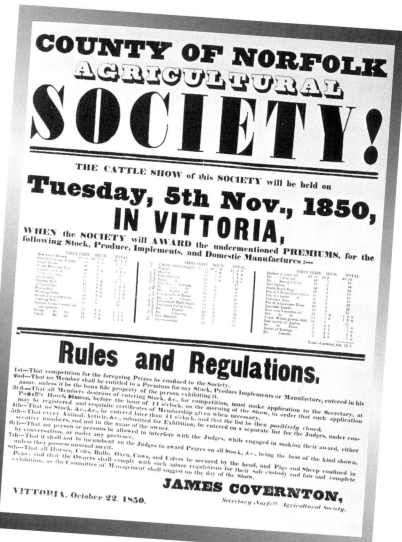

Poster advertising the County of Norfolk Fair in 1850. Prizes offered and rules are printed right on the poster. This fair still operates today in Simcoe, Ontario.

Agricultural Societies in Ontario

The new province of Upper Canada was created in 1791. The first lieutenant governor, John Graves Simcoe, created an elected legislative assembly and an agricultural society in the same month—July 1792. The all-encompassing Agricultural Society of Upper Canada held biannual meetings at Niagara-on-the-Lake. In true English aristocratic tradition, prominent men of the new colony met over supper to discuss agricultural issues. A small library of books on agriculture was created. But few practical or "real" farmers subscribed to the society. Attempts to hold a fair at Queenston in the late 1790s failed. Distances were too great, travel too difficult, and Ontario was just not ready for agricultural societies at this stage in its history. Various district fairs, livestock markets and agricultural co-ops were attempted during the first three decades of the nineteenth century. By 1830, conditions had changed, with many areas of the province sufficiently settled to allow farmers to organize agricultural societies. In that year, the colonial administration of Upper Canada, realizing the value of agricultural societies, passed an act granting £100 to any district that first raised £50 in subscriptions or memberships. In a few years, all eleven districts in Upper Canada had formed agricultural societies. County and township societies began to appear, and the grant total grew to a princely £1,607. The act stated that the agricultural societies would be given grants for the purpose of "the importing of valuable livestock, grain, grass seeds, useful implements of husbandry or what ever else might induce the improvement of Agriculture."[6] No mention was made of holding fairs, exhibitions or markets. These early societies dabbled in such areas as promoting agricultural implements, importing gypsum (used as lime), agricultural newspapers and books, and livestock. The Prince Edward County Society, based at Picton, kept two bulls on inventory, the big one for use of its members, the small one for rental to non-members. The fee: 5 shillings per cow. Membership had its privileges!

Despite their many other activities, agricultural societies in Ontario also organized fairs. The Midland District Fair held at Napanee in 1836 included the following: 108 horses, 100 cows, 47 oxen, 40 calves and 1,200 bushels of grain. All these items were to be sold at the event. The prize list for the Frontenac Agricultural Society fair held on October 12, 1825, at Kingston offered the following classes:

Livestock	*Grain*
Best bull	Best 2 acres of wheat
Best boar	Best 2 acres of oats
Best sow	Best acre of corn
Best yoke of oxen	Best 2 acres of peas
Best ram	Best acre of barley
Best 6 ewes	Best acre of potatoes
Best three-year-old steers/heifers	Best half acre of flax
Best cow	Best quarter acre hemp

Miscellaneous	
Best plowing with horses	Bests 20 yds of flannel
Best plowing with oxen	Best 20 yds of linen
Best maple sugar	Best cheese[7]

While the show ring and the exhibit hall were both used, standing field crops and plowing matches were also included in the show.

Many of the earliest settled regions of the province had formed agricultural societies by 1850. The pre-1850 societies included:

Williamstown 1812
Kingston 1825
Delta 1830
Lindsay 1833
Wellington County 1837
Richmond 1836
Simcoe County 1838
Whitby 1838
Merrickville 1838
Norfolk County 1840
Puslinch (Aberfoyle) 1840
Stratford 1841
Carleton County 1844
Van Kleek Hill 1845
Esquesing (Brampton) 1847
Woodbridge 1847
Thorold 1847
South Perth 1847
Ingersoll 1848

By 1850, the organizational representation of the agricultural society was changed from the district, which was too large, to the county/township. Each township was encouraged to form an agricultural society. It was reasoned that a township should contain between five hundred and one thousand farmers, and this number was considered suitable for an agricultural society. A county consisted of between six and twelve townships. The county societies were expected to hold regional exhibitions at which local township winners could compete to advance to the next level of competition. An even higher level of competition was reached at the provincial exhibition. The county shows were to be concentrated in larger towns or cities, usually the county seat. As Ontario's population grew, the county system was modified to the electoral division society. Electoral divisions were based on population, and each electoral division sent one member to the Ontario Legislative Assembly. Some of the more populous counties such as York, Niagara, Wellington or Grey, might have two or three electoral division societies or county fairs. Grey County is a good example. The original Grey County Agricultural Society was divided into the Grey North (Owen Sound), Grey South (Glenelg) and Grey East (Flesherton) electoral agricultural societies. While the early county/electoral societies tended to rotate their fairs from place to place (politics is wonderful!), the exhibitions

A typical turn-of-the-century fairgrounds, Kinmount Fair, Ontario, circa 1900. A racetrack and grandstand testify to the importance of horse racing at the fair. A board fence enclosed the grounds. The large tent housed a carousel or a merry-go-round, the only ride available to a smaller fair at this time. As the background testifies, this area of Ontario was not "prime agricultural ground." But that did not stop local residents from organizing a fair. Agricultural fairs were so Canadian, they were found everywhere, even in the Yukon!

eventually became settled in one centre. Eventually the county/townships system died out, and while fairs were often classified by size, a fair became just a fair.

Ontario, or Canada West, became fair crazy in the latter half of the nineteenth century. Every township and county had to have a fair: it was a mark of prestige. As settlement spread north and west, the new settlements immediately set up their own agricultural societies. In the 1860s, the Ottawa-Huron Tract (the land between Georgian Bay and the Ottawa River) was opened for settlement. In 1860, the newly surveyed township of Dysart, in Haliburton County, contained one roofless shanty. Yet in 1865, the first Dysart Fair (Haliburton Fair) was held in Haliburton Village, despite the fact that the whole township contained only 28 families and a mere 55 cleared acres. This smallish fair was described as follows:

"The exhibits were displayed before Erskines' blacksmith shop with the ladies' work shown to advantage on a fine line stretched above a carpenter's workbench, the top of which held two pairs of ox-bows and a complete ox yoke with wooden boat keys all constructed by local artisans. Under the workbench were turnips, potatoes and other garden produce including two fine pumpkins and some corn grown by Mr. David Sawyer."[8]

The Haliburton Fair continued into the 1930s. By this time agriculture no longer figured in the local economy. The community had changed, and the fair disappeared from the village. It amalgamated with the neighbouring Minden Township Fair.

The last frontier in Ontario was in the northern section of the province, particularly the clay belts along the Quebec border. Around 1900, the government of Ontario set up a five-point plan to encourage settlement in Northern Ontario. This plan called for public funds to support:

1. Colonization roads
2. Schools
3. Police
4. Welfare
5. Agricultural societies

Agricultural societies were to be on the leading edge of government policy. Their usefulness was well established by 1900. It was hoped these new agricultural societies would lead to an agricultural revolution in Northern Ontario.

By 1900, the role of agricultural societies had changed. The following chart reflects the nature of this change:

	1854		1878	
	County	Township	County	Township
Number of societies	41	175	88	286
Exhibitions held	41	152	84	253
Livestock for use of members	11	42	1	12
Sold seeds, grain, etc.	12	18	2	11
Imported livestock for sale	4	5	3	6
Implements for sale	2	4	1	0
Prizes for field crop competition	4	7	0	5
Periodicals given to members	14	41	4	15
Held plowing matches	5	26	16	22[9]

Most societies held some sort of exhibition or fair each year. Less than 10 per cent did not hold a fair of their own. Many of these societies held "union" fairs with other societies. In some counties, county/electoral societies would join with the township society to hold a joint fair. It was just too impractical to hold two fairs in one town every year. For example, the South Electoral Division of Victoria County and the Ops Township Societies both wanted Lindsay as their fair site. A union fair was the solution. With the exception of plowing matches, all other agricultural society activities declined during this era. By the early 1900s, agricultural societies were left with fairs as their primary activity.

By 1914, almost 400 fairs were held every year in Ontario. Since this high-water mark, the numbers have declined, and today, there are 230 fairs throughout the province. The decline in the farm population, improved transportation, the growth of urban centres and general changes in society have reduced the number of fairs in Ontario. Many societies amalgamated; some just disappeared. Ontario fairs were originally concentrated in a six-to-eight-week span between Labour Day and Thanksgiving. This meant forty to fifty fairs were held each week in Ontario. Eventually, competition for exhibitors, midways and crowds led to a "spreading out" of the fair season, and summer fairs, in July and August, became common. The fair season now starts in May, although the traditional season, September and October, still remains popular.

Agricultural Societies in Western Canada

Winnipeg Exhibition, 1909. In the foreground is a picnic area complete with boardwalk. In the background, the grandstand is festooned with advertising for the Hudson's Bay Company stores. The Winnipeg Fair was discontinued in 1917, and for many years the "Gateway to the Prairies" was without an agricultural fair.

By the late 1800s, agricultural fairs were well established in the eastern provinces of Canada. Most agricultural societies had abandoned their efforts in livestock management and machinery sales and instead, concentrated on fairs. The fair system was soundly entrenched in the longer-settled regions of Canada. As new frontier areas opened up, the pioneers brought their love of fairs with them. In 1869, the new Dominion of Canada purchased the vast Hudson's Bay Company territories. The new owners declared the Prairies, called the Northwest Territories, open for agricultural settlement. It is no coincidence the first Prairie fairs began shortly thereafter. The first order of business in new communities in Western Canada were:

1. A school
2. A church
3. A fair

All three could happen simultaneously — and often did.

In 1871, the hamlet of Winnipeg (population 500) staged a small fair. Hardly had the fair opened when a messenger arrived with the fateful news that a Fenian raiding party had crossed the Manitoba border. The Fenians were Irish-Americans who tried to seize Canada and hold it for ransom as part of their fight for Ireland's independence. In exchange for getting their colonies back, the British were to withdraw from Ireland. It was a hare-brained scheme, but it caused a lot of panic in

Threshing Machine Exhibit, Brandon Fair, 1904. The machinery dealers have their newest models ready for the throngs of curious farmers. The best advertising was always at the fair.

British North America. News of a possible Fenian raid threw the fair-goers into a panic. The fair was immediately cancelled, and everyone headed home to arm themselves or hide. It was to be another twenty years before Winnipeg attempted its next fair.

The honour of holding the first continually held fair in Manitoba fell to Portage la Prairie in 1872. The first president, an ex-Ontarian, demonstrated his enthusiasm by entering every class. A banquet closed the fair, but the president cancelled the after-dinner speeches. He had cows to take home and milk. By 1882, there were nineteen agricultural societies in Manitoba and seventeen of them held fairs.

In 1874, the federal government made $10,000 in seed wheat available for farmers devastated by a recent grasshopper plague. The agricultural societies of Manitoba were called upon to handle distribution of this seed grain: a sign of the high esteem given these organizations in the West.

Saskatoon Fair, 1910. At left, two exhibit halls, one with a cupola, a crude attempt to imitate a Crystal Palace. At right, the midway and its sideshow tents. In the centre is St. John's ambulance first-aid station. At the front centre, a small drive-in restaurant, with a one-horse-buggy lane, a fast-food harbinger! The grounds are enclosed with a board fence, which serves as a handy horse-hitching post. Beyond the fence, the lone and level prairie stretches out of sight.

ANNUAL EXHIBITION

OF THE

LORNE AGRICULTURAL SOCIETY,

PRINCE ALBERT,

Tuesday, August 5th, 1902.

PROGRAMME.

Horse Races.

	1st	2nd	3rd
Half Mile Pony Race, under 14½ hands, 2 in 3, purse $20.00	$10.00	$6.00	$4.00
Farmers Trot or Pace, to be owned and driven by farmer; purse $20.00	10.00	6.00	4.00
Half Mile Indian Pony Race, 2 in 3, to be owned and riden by Indian	6.00	4.00	2.00
Half Mile Scurry; start 100 yards behind grand stand, ride to grand stand bareback, saddle up, and ride once round track	5.00	3.00	2.00
Cigar and Umbrella Race; saddle pony, light cigar, put up umbrella, and mount. First round course with cigar alight and umbrella up, wins	5.00	3.00	2.00
Brandy and Soda Race; start dismounted, saddle your horse and ride to a table on which will be found tumblers and bottles of soda water; open the soda water and pour contents into the tumblers, drink and ride home	5.00	3.00	2.00

Football—Society's Cup—Valued at $25.00. If more than two clubs compete the Clubs will play during the forenoon, except the final draw, which will be played at 4 o'clock p. m. Games to be played under rules and regulations of last year.

Miscellaneous Sports.

	1st	2nd	3rd
Quoit Throwing	$2.00	$1.00	
Trap Shooting, Open, shoot at 20 birds.	5.00	3.00	$2.00
Bicycle Costume Race; mount without assistance, at the word go, ride to flag, dismount and open parcel bearing your number, dress yourself in the costume found, then remount and ride home round course	3.00	2.00	1.00

Athletic Sports.

	1st	2nd	3rd
Running Long Jump	$2.00	$1.00	
Running High Jump	2.00	1.00	
Running Hop Step and Jump	2.00	1.00	
Vaulting with Pole	2.00	1.00	
Putting Stone	2.00	1.00	
Relay Race, one mile, eight on each side; contesting parties run in pairs, carrying a stick, which is passed on to the next pair, stationed 220 yards distant. Contesting parties run in opposite directions	5.00		
Men's 100 yard Foot Race	3.00	2.00	$1.00
Boys 200 yard Foot Race under 14	2.00	1.00	50c
Girls 100 yard Foot Race under 14	1.00	50c	25c
Squaw Race	2.00	1.00	
Fat Mans' Race, 200 pounds or over	Box Cigars.		

RULES and REGULATIONS—Horse Racing.

Four to enter, three to start. Entrance fee 10 per cent. of purse, except Indian pony race, entry free. A horse distancing the field entitled to first money only. The committee reserve the right to declare off or call out any horse anytime after 2 p. m., Tuesday, 5th August, or postpone or do any other action which they may think advisable.

Admission to Grounds—25c.   Children under 14—10c.

Admission to Grand Stand—15c.   Children under 14—10c.

W. C. McKAY, President.   W. J. KERNAGHAN, Secretary.

In May 1881, the Canadian Pacific Railway (CPR) arrived at the present site of Brandon, Manitoba, which consisted of a shanty, a well and clothesline. One year later, the town was incorporated, and four months after incorporation, the first Brandon Fair was held. Now, that's quick growth!

For years, Brandon and Winnipeg carried on a friendly rivalry for the honour of calling their fair the Manitoba Provincial Exhibition. The Winnipeg Industrial Exhibition was cancelled in 1915, and Brandon won the title permanently.

Brandon became famous for its livestock, especially draft horses. Most new settlers to the Prairies were funnelled through Winnipeg. Most of them needed to purchase livestock, especially horses, and the ideal place to acquire these necessities was as close to the potential homestead as possible. Thus, Winnipeg was filled with homesteaders trying to purchase horses and other farm animals. The more settled regions of Manitoba were only too glad to produce them as a "cash crop." Soon many Manitoba farmers were specializing in purebred livestock and horses. For many years the most competitive class at any Prairie fair was the Clydesdale stallion class at Brandon Fair.

The number of Manitoba agricultural societies peaked in 1930 when eighty-two fairs were held. This number collapsed to as low as twenty-five fairs during the Depression years. By 1970, there were three "A" fairs (Brandon, Red River Exhibition in Winnipeg, Brandon Winter Fair), six "B" fairs and fifty-four "C" fairs, for a total of sixty-three.

As the flood of settlers gushed westward in and beyond Manitoba, agricultural societies began to appear in the Northwest Territories, now Saskatchewan and Alberta. The Battlefords Agricultural Society was established in 1883, but its first fair was delayed until 1887 due to the North-West Rebellion, also called the Riel Rebellion. It was hard to hold a fair when the town was under siege. By 1884, Regina, Prince Albert, Moose Jaw, Indian Head and Whitewood had fairs. Carlyle, Qu'Appelle, Yorkton

Steer roping, territorial exhibition, Regina, 1895. Rodeo was still in its infancy. The ring was too large, and take-downs were often made out of sight of the spectators. Here, steers for the next act are patiently waiting their turn.

and Wolsely followed in 1885, and Saskatoon and Grenfell the next year.

Not all early fairs were exemplary successes. Take for instance Cannington, an early settlement of English aristocrats with rather eccentric tastes. At the first Cannington Fair, a bagpiper was brought in for the princely sum of one bottle of whiskey. So frightening was the wail of bagpipes to the local livestock, they bolted the grounds causing the fair to end prematurely as the fair-goers rushed to round them up. Maybe the whiskey had been consumed before the show instead of after.

In 1895, the Legislative Assembly of the Northwest Territories decided to hold a territorial exhibition. Regina was selected as the site. Prize money exceeded $19,000 and the show was billed as the biggest show west of Toronto. The fair featured classes for buggies, farm gates, pumps, saddles, bricks, buffalo coats, brooms, cigars, straw hats and soap: everything needed in a pioneer community. One saddle-horse class was entered under the rule that all entries became property of the North-West Mounted Police for $125 (for the horse, that is, not the riders). To encourage mixed farming, one class called for an entry to consist of wheat, oats, barley, peas, flax, cattle, sheep and pigs. Another award of $500 was given to the best prairie-fire extinguisher; to consist of no more than four horses and two men! Now, that's a demonstration I would like to see. Regina was so exhausted after the territorial exhibition that the next fair was not held until 1899. At the 1903 Regina Fair, admission charges included 15¢ for one-horse rigs and 25¢ for two-horse rigs. Rigs carrying exhibits got in free. Pies seemed to be the most popular exhibit passing through the gates. Safe to say, most pies never made it to the exhibit hall!

The number of fairs in Saskatchewan mushroomed from 33 in 1905, to 110 in 1915, to 158 in 1930. The Depression hit hard, reducing the number of fairs to a mere twenty-nine in 1933. Many agricultural societies went dormant and resurfaced later. By 1965, the number of societies was seventy-four, with fifty-four fairs held that year.

To Edmonton belongs the honour of staging the first fair in the

The first fair held at Three Hills, Alberta. The community had just been founded, and featured only a store and a post office, with flags. There are no plowed or fenced fields, just open prairie. Clearly this was cattle country, but before the farmers fenced their plots, built barns or proper roads, they held a fair. Priorities are priorities.

Innisfail, near Calgary, was a typical prairie town. Railway station, water tower, windmills and fairgrounds. A saddle-horse class is the attention in the show ring. Conditions on the grounds were still primitive, as the tethered bull in the foreground shows. That fence would never have held him if he wanted to go wandering.

Northwest Territories in 1879. At the time, Fort Edmonton was a fur-trading post still heavily under the influence of the Hudson's Bay Company. The HBC encouraged agriculture in the region by donating cash, prizes and buildings for the first fair. The first show was very successful, but the second fair was somewhat less spectacular as no directors bothered to show up! A class for pemmican was held, but unfortunately the entries of dried buffalo meat were devoured by dogs before judging could take place. Such events always made the city of Edmonton concerned about its image as just a fur-trading town. In 1912, in an attempt to shed this rough image, the directors of the Edmonton Fair banned rodeos, auto races and "aeroplane"-stunt shows. Horse racing was also briefly banned in 1918. However, the banned shows were quickly reinstated — and the fun continued.

The early Edmonton Fairs were so successful that other fairs sprang up all over the province. A 1903 fair schedule for what is now Alberta included the following fairs:

Edmonton, June 30–July 2
Wetaskiwin, July 3–4
Calgary, July 7–10
Strathcona, August 13–14
Fort Saskatchewan, August 17–18
Lacombe, August 20
Olds, October 6
Innisfail, October 7
Red Deer, October 8–9
Vegreville, October 9–10

The Calgary and District Agricultural Society was formed in 1884. The Calgary district was cattle country. In 1886, 500 people lived in the village of Calgary, which was a railroad station on the new CPR line, but 75,000 cows grazed along the Bow River Valley. Instead of holding a fair, a rail car showing exhibits from the area was sent back East. The first actual Calgary Fair, held on October 18, 1886, was snowed out!

The first fair held at Griffin Creek, Alberta, in 1914. Facilities were rudimentary: a pole barn, a few tents and some tie-ups for livestock. No large midways or exotic sideshows: just a plain old-fashioned agricultural fair. In the foreground, two fair-goers inspect a cow, a sight that makes an agricultural purist smile.

Needless to say, the date was changed. A terrible weather year in 1895 destroyed crops and led to the fair being cancelled. It took five years for the fair to resurrect itself. The Calgary Fair received a big boost in 1908 when it was awarded the title of Dominion Exhibition. The federal government granted $50,000, the new province of Alberta $35,000 and the city of Calgary $25,000. Attendance exceeded 100,000, with most of the visitors travelling by rail. The first Stampede was held in 1912, but only after 1919 did it become a permanent event. In 1923, the Stampede and the Fair were amalgamated.

The Saskatoon area began as a "temperance colony." An advertising display seeking new settlers for the area appeared at the CNE as early as 1881

The residents of Calgary lived the Stampede. To add a little zest to fair week, these cowboys have ridden their horses into the store to shop from horseback. Evidently it was a crowd pleaser. One cowboy rode his faithful steed into the "Edmonton Club." He was asked by the management to apologize for his actions. He apologized ... to his horse, for exposing the faithful steed to such company.

with an agricultural society, the Temperance Colony Agricultural Society, being organized in 1884. One of the first acts of the newly incorporated town of Saskatoon was to pass a bylaw granting $30,000 for a new fairgrounds. This was a fantastic sum for its time, and it reflects the commitment that the early Prairie pioneers made to their agricultural societies and fairs.[10]

Agricultural Societies in British Columbia

Victoria hosted the first Pacific agricultural fair in 1861. The colony contained only three hundred farms, all on Vancouver Island. Most of the early influences on British Columbia fairs came directly from Britain, as the isolated colony had little contact with the other British North American colonies. New Westminster had a fair by 1865, followed by Cowichan, Saanich in 1868, Chilliwack, Surrey, Richmond, Courtney, Mission, Kamloops, and Armstrong.

Sir Wilfrid Laurier opened the first Pacific National Exhibition (PNE) in Vancouver in 1910. At the first PNE, cattle were tethered to Douglas fir trees, a lovely image. To raise funds, lifetime memberships were sold, and $20,000 was realized to help get the new fair started. By the ninth fair, attendance had topped 100,000. In 1927, 280,000 people passed through the gates of the PNE. It was estimated that 60 per cent of all British Columbia farmers went to the PNE each year. A winter fair was added in 1925. In a pattern repeated all across Canada, the PNE grounds became the centre for skating, curling, golf, concerts, circuses, horse racing and more.

The main goal of the first PNE was to promote industry in the growing city of Vancouver. New Westminster was the home of the local agricultural fair, with Vancouver holding the industrial exhibition. A state of uneasy competition between the two cities lasted for twenty years, until a merger in the 1930s, due to a fire that destroyed the buildings at New Westminster. Vancouver had a magnificent facility at Hastings Park, and many diverse events were held here over the years. In fact, the Hastings Park fairgrounds have played a big role in Vancouver history and culture. When Vancouver was awarded the Commonwealth Games in 1954, Hastings Park became the main site. Then the newly built Commonwealth Stadium helped Vancouver secure a Canadian Football League franchise in 1964. In 1970, the newly minted Vancouver Canucks of the National Hockey League called Hastings Park Coliseum their home. The

The British Empire and Commonwealth Games, later named the Commonwealth Games, were opened July 30, 1954 at Hastings Park, site of the Pacific National Exhibition.

Rodeos spread from Western Canada all over the nation. Williams Lake, B.C., was a centre for cattle grazing and hence prime territory for rodeos. By this date (1931), most patrons arrived by car, but a few horse-and-buggy conveyances are still evident. The concession area is small, and is clustered in someone's barnyard. But they do have a racetrack!

Yukon Horticultural and Industrial Fair, Dawson City, September 9–12, 1903. Wow! The fair circuit extended to the Yukon as early as 1903. The agricultural community was virtually nonexistent, but that didn't stop the fair spirit. Food products seem to be the predominant merchandise. Flags of many nations decorate the hall. Dawson City was then on the downside of the famous gold rush of 1898, but nowhere do we see the evidence of sourdoughs, gold mining or a rough-and-tumble frontier town.

53

Stanley Cup and the Grey Cup have both been here. The story of this multi-use fairgrounds is typical of communities throughout Canada. The fairgrounds has been part and parcel of community culture and folklore.[11]

A family scene, Mission Fair, B.C. Many families were heavily involved at the fair. This family is posing with their roles in the fair. These range from showing livestock to playing in the band.

Victoria, B.C. fairgrounds, 1902. In the background is an elaborate crystal palace. In the foreground, another common fair feature: the race track!

The number of fairs in Canada peaked in 1920, not long after World War I. After this date, the number of fairs went into a slow decline, and very few new fairs were established. There were many reasons for this. By 1920, most of Canada was "settled." The main agricultural areas of the country had been opened up, and very few new agricultural frontiers awaited the farmer. After this date, coincidentally the number of farms and farmers also began to decline, a pattern that accelerated during the Great Depression and later, during the urbanization of many areas of the country. Changing conditions led to a drop in the number of fairs in Canada over the next century. This was not necessarily a bad trend. The existing fairs grew larger and stronger. The number of Canadians patronizing fairs actually increased as the number of fairs shrank.

As Canada changed, fairs changed too and cemented their role in Canadian society. They evolved and adapted. Most fairs were flexible enough to roll with the times. The story of agricultural fairs is the story of change, and yet, the more they changed, the more they stayed the same. It has always been a fine line for Canadian fairs: to meld past and present, urban and rural, education and entertainment, old and new. But they have done it!

PROVINCIAL EXHIBITIONS: THE ALLSTAR FAIRS

At one time or another, all provinces in Canada have held a provincial exhibition. Provincial exhibitions were always considered province-wide fairs, the main goal to attract the best exhibits and livestock from all over the province so that ideas and livestock could be readily exchanged. Early attempts before 1860 failed for a variety of reasons. These super-fairs needed a good circuit of regional and local fairs to start up the exhibitors and spur interest. Distance and transportation difficulties were also inhibiting factors. But once the fair circuits became more organized in each province, the idea of a provincial exhibition became practical. It became a badge of honour for a local fair to be granted the privilege of organizing a provincial exhibition. In fact, competition for these exhibitions was so fierce that rival cities engaged in outright bidding and lobbying for the right to hold a provincial exhibition. It was a mark of honour, a coming of age and a signal that a city had reached an important stage in its development to be awarded the title "provincial exhibition."

By the 1840s, Ontario, the pioneer of provincial exhibitions, was developing a thriving system of local fairs, a rudimentary transportation network and a need to hold a year-end show. The Agricultural Association of Upper Canada planned to hold the first provincial exhibition in 1846. Toronto was chosen as the site because it was the largest centre in the former Upper Canada colony, it was fairly easy to get to, and most of the directors were residents of that city. Four hundred pounds in prize money were offered with 10 for the best stallions and 7.5 for the best bull, 5 for the best ram, and 5 for the best boar. There were 1,100 entries, and the fair was well attended, making this first exhibition a triumph. A big factor in its success was the participation of the steamship companies, which ran special reduced-fare excursions to the exhibition. The event lasted for four days, and in true Berkshire fashion, ended with a plowing match and banquet. Everyone—directors, exhibitors and fair-goers—were pleased with the event. As a result, plans were laid for future events of a similar nature.[1]

Since the provincial exhibition was to be for all Ontario, future events were to be rotated among the major centres. In an era of limited means of transportation, it was only right that different sections of the colony got their chance to host the fair. As a result, a circuit gradually evolved for the provincial exhibition. And therein lies the downfall of the provincial fairs system, both in Ontario and across Canada: they were so successful that each city wanted to hold one every year. To wait four or five years for another exhibition was intolerable. Gradually, the host cities left the circuit and established their own exhibitions. London, Ontario, formed the Western Fair in 1871 following its last provincial exhibition, and Toronto started the CNE in 1879, the year after hosting its last provincial show.

Ontario was filled with large fairs by 1886 as each region staged its own exhibition. The need for a province-wide super-fair was disappearing as super-fairs were everywhere. The last Ontario provincial exhibition was held in Guelph in 1889.

The concept of a provincial exhibition was transplanted to Western Canada with the early settlers. Fledgling fairs vied for the title "provincial exhibition." Brandon and Winnipeg competed for the title "Provincial Exhibition of Manitoba" as early as the 1890s. Brandon Fair still proudly uses that title today. A Northwest Territories Exhibition was held at Regina in 1895, before Saskatchewan and Alberta existed as provinces. The exhibition was a symbol that the town of Regina was emerging as the leading centre in Saskatchewan.

The title "provincial exhibition" was a source of conflict between Calgary and Edmonton as well. British Columbia had a provincial exhibition based in Victoria and later at New Westminster. When New Westminster cancelled its fair in 1930, the title was transferred to its rival in Vancouver. Unable to compete with its larger rivals on the coast, Armstrong fair in the Okanagan Valley adopted the title "interior provincial exhibition."

So successful was the provincial-exhibition concept that the federal government set up a Dominion Exhibition circuit on a national scale. Ottawa hosted the first Dominion Exhibition in 1889. Thereafter the Dominion Exhibitions were rotated around the country, with the sites selected based on geographical and political considerations. The status of Dominion Exhibition also carried some healthy grants with it. In 1909, Calgary received $50,000 from the Dominion government to stage the Dominion Ex. The new province of Alberta matched this sum and the city of Calgary contributed as well. All this coin built some nice new facilities and left a lasting legacy for Calgary. The title "Dominion Exhibition" carried a lot of prestige in its heyday, before it was cancelled in 1914.

RED FIFE WHEAT: A CANADIAN SUCCESS STORY

Nothing symbolizes the role agricultural societies played in agricultural improvement more than the story of Red Fife wheat. The tale starts innocently enough on the farm of David Fife near Peterborough, Ontario. Fife was having a problem finding an acceptable variety of wheat to grow on his farm. Short growing seasons, diseases and severe winters were causing grief with the wheat varieties the early settlers had brought from Europe. Fife complained to a Scottish friend in 1840 about his wheat problems. The friend acquired some wheat for a Russian shipment and sent the samples to Fife. Legend has it, the friend scooped up a handful of grain from a Russian vessel and hid it in his hat. Most of the samples were unsuitable, but five stocks seemed to fare better than the rest. They were resistant to rust disease and matured earlier. A wandering cow ate four stocks, but Mrs. Fife saved the remaining one. The third year his crop was 1/2 bushel, and Fife realized he was on to a good thing and had his neighbours try the new variety. The results were most encouraging. At this stage, David Fife did the natural thing for his time and era: he approached the local agricultural society.

In 1849, the Otonabee Township Agricultural Society purchased 250 bushels of Red Fife wheat for $2 a bushel and sold the seed to its members. The results were gratifying. The secretary of the society wrote a letter to the Agricultural Association of Canada West (as Ontario was called before 1867) and the *Ontario Agriculturalist*, a farm journal. The article attracted attention and buyers from all over North America. Agricultural societies all across Ontario began to distribute Red Fife seed wheat to its members and heartily endorsed this new breed of spring wheat. By 1860, Red Fife was the premiere spring wheat variety in Canada and the northern U.S. In the 1870s, a group of Manitoba farmers secured a load from the U.S., a delicious irony as it was a Canadian product. The wheat loved prairie soil and prairie climate. In 1883, Manitoba agricultural societies restricted all entries of spring wheat to Red Fife only. A message was sent!

In 1876, 857 bushels of Red Fife wheat from Manitoba arrived back in Toronto, the first drop in an ocean of wheat that would soon flow from the Prairies. The quality was impressive, and in the next few decades, the quantity would also become impressive. Agricultural researchers at various experimental stations improved upon the original Red Fife, first with the Marquis offshoot and since then with countless other brands all descended from that single stalk. Wheat quickly became the predominant cash crop from Western Canada. After 1900, it was Canada's number one export. Red Fife led Canada into the forefront of grain-producing nations. And it all started with one stalk and a system of agricultural improvement societies dedicated to advancing agriculture as an industry.

The Midway: Carousels and Conmen

OF ALL THE SEGMENTS OF A FAIR, the midway or carnival is the one that garners the most attention. It is beloved by some and cursed by others. It is both a source of joy and a source of disgust. It attracts and repels. And yet it is an undeniable part of every fair. Or maybe I should say most fairs, for there are still a few holdouts that refuse to include a midway or cannot find one willing to come to their fair! Yet midways and fairs are firmly bound together by history and by economics. To quote Joe McKennon, carnival historian:

"There is not a large carnival in North America today that could not exist without a fairly substantial route of so-called agricultural fairs."[1]

The history of midways/carnivals is part of the history of fairs and exhibitions and vice versa.

Ye Olde Carnival History

The ancestors of midways/carnivals actually existed before those of agricultural fairs. Sideshows, entertainers and hucksters were part of fairs in

A midway sideshow, Johnny J. Jones midway, Edmonton Fair. The human-mermaid show mixed a little athletics, myth and sex all in the same tank. Here, the elaborately painted fronts look enticing enough, but appearances by the actual mermaids have drawn only a small crowd of gawkers, mostly male, of course!

Another exotic sideshow from the Johnny J. Jones Midway. The show must not have been too risqué, for children were admitted for 10¢ (adults, 20¢). Not many fair-goers seem enticed but the two ticket sellers seem to hint at expectations of much larger crowds.

biblical and medieval times. Agriculture came into the mix in the 1700s. Wherever crowds gathered, so did the carnival types. Festivals and trade fairs had an inate carnival/midway element. The earliest carnivals consisted of buskers, freak shows, animal shows, plays, musicians and games of chance. Mechanical riding devices came along later. The French pioneered the carousel or merry-go-round in the 1700s. The Ferris wheel dates from the 1890s, while the jerking rides, like the whip, appeared after 1900. Midways/carnivals have changed greatly in the last hundred years. Some of these changes are due to changes in technology; some are due to changes in society. The story of the midway/carnival is also a story of change.

The earliest English trade fairs, such as St. Bartholomew's and Sturbridge, always had a carnival element. Buskers, actors, conmen and circus-style performers were a big, and often unwanted, element of these medieval fairs. The earliest North American fairs were without carnivals. While agricultural purists wanted it that way, the fact was, carnival acts were scarce or nonexistent in pioneer North America, and the early fair organizers worked with purely agricultural shows. But this was not to last. In 1825, the first carousel arrived in North America via England. By the mid 1850s, animal circuses became commonplace. P.T. Barnum began to tour North America with his "Big-Top Circus." Barnum's early circus, transported by horse-drawn vehicles, was very limited in its travelling range. The advent of the steam train made circus tours more practical. By 1872, Barnum was using fifteen railway cars to move his circus.

Midway scene. In front is the famous "Over the Falls" show, flanked by sundry freak/geek shows. One gets curious about just what is inside. In the background are "gilly cars," on which the shows were loaded and dispatched to the train station for the journey to the next town. Gilly cars were placed on flatcars and simply tied down.

In the same year, 37 circuses were on tour throughout North America.[2] But while agricultural fairs might have a circus as part of their lineup, they remained separate and seldom worked with the touring circuses. Distinct from, though related to, circuses, a whole series of smaller acts, shows and buskers began to appear and began to tour fairs and festivals as independents. Only after 1893 did carnival operators begin to unite these independent acts into organized midway/carnival companies.

The great Columbian World's Fair, held in Chicago in 1893, was the catalyst for the midway/carnival industry. So lucrative was this exhibition that carnival and sideshow operators began to ponder joining together to form organized companies and going "on tour."

The first primitive midway company hit the road for the 1894 fair season. But the tour was a total bust and the midway was dispersed before season's end. It was just too difficult to find profitable shows to play, especially early in the season. The midway operators and the fair managers were not in sync at this early stage in midway history. A similar attempt the next summer also ended in disaster. But the carnival operators still remembered the success of the Chicago Midway Plaisance of 1893, and kept trying. In 1896, three Englishmen, Frank Bostwick and the two Frerari Brothers landed in New York with an English-style midway. This included animal acts, sideshows and several riding

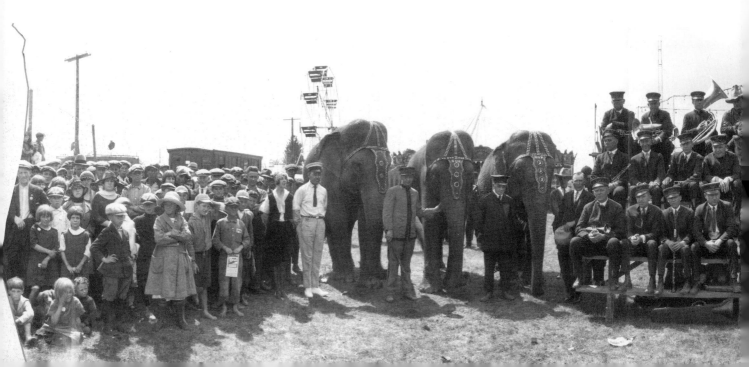

devices. The three began to tour a circuit in the traditional fair season and also performed in amusement parks in the off-season. In 1898, another group operating out of Chicago began to play at street festivals and fairs. The Chicago Midway, as it was known, was sponsored by the Elks lodges, a community-service organization. These lodges became involved in street festivals as fundraisers. So successful were these street festivals, several prominent Elks organized their own midway/carnival groups and took them on the road to play both street festivals and agricultural fairs.

For the 1900 season, there were three major midway companies on the road. An ad for one, the Bostwick-Frerari Carnival, listed 14 attractions:

Frerari's Wild Animal Arena
Pro. Woodward's Dog and Monkey Circus
Streets of Cairo (with 5 camels)
Electric Theatre
Trip to the Moon
German Village
Chameleon Lady
Electra and Prismatic Fantasia
Edison's Animated Pictures
King Dado Snake Eater
Conderman Ferris Wheel
Turkish Theatre
$25,000 Crystal Maze
Original Wild Girl

Two years later, 24 midway companies hit the road for the 1902 fair season.[3]

There were many advantages to having a formal company structure for the midway/carnivals. By uniting the various acts and rides, the company could attract more patrons. The midway could also plan a set route

Johnny J. Jones Midway, Edmonton Fair, 1924. The entire cast poses for a group panorama. It takes a lot of different people to make a large midway run. The costumes betray their role or status. See if you can pick out the different acts.

of show dates, thus having the security of guaranteed work. The larger midway companies could also negotiate deals with more and larger fairs. This in turn helped both the fair and the midway/carnival company. Transportation was also easier to organize, especially with railways. As riding devices became more prevalent, the heavy cost was more easily absorbed by larger units. Safety and support were easier in larger numbers as well. The midway/carnival companies often merged to form more economically viable units. Bigger shows with better and bigger attractions were more likely to win the right to play at the larger and more profitable fairs.

To organize a carnival for a year's tour, a midway owner had to do two things: line up the acts and line up the route. Acts could be acquired by contacts within the industry, but most carnival operators advertised in *Billboard* magazine for their shows. *Billboard*, based in New York City, was the official voice of the carnival/amusement/midway industry in North America. The business was a North American-wide industry, with most of the midways/acts coming from the U.S., but playing both Canadian and American fairs. It was truly an international business. In the first half of the twentieth century, it was free trade in the carnival business, or "midways sans frontiers."

The carnival season usually ran from May to October. Since agricultural fairs were only held from August to October, midway operators tried to fill in the remaining weeks with "still dates." A still date was a festival or carnival, usually much smaller and of shorter duration than a fair. Good still dates were hard to find and most were only one-day events. The midway companies found it hard to recoup their expenses in the three to four hours these festivals operated. Many midways limped into the traditional fair season deeply in the red. As a result, midway operators really looked forward to the opening of the agricultural fair season.

In the winter off-season, midway/carnival operators planned their

routes. They attempted to sign a contract, or play one event, each weekend of the season. The weekends were the key, for most events were held at the end of the week. It was also very difficult to pack up and move the carnivals twice in a seven-day span. It was much easier to set up and knock down only once a week. A great deal of labour was necessary to move locations. With the advent of mechanical rides, even more effort was expended on each move.

Pictured on this page is the route list for carnival operators Conklin and Garrett for the 1932 season. It gives some idea of the gruelling schedule a midway company faced. Remember, it was the height of the Great Depression and money was scarce. The midway covered a lot of miles (from Vancouver, B.C. to Granby, Quebec) and didn't have a lot of time between shows to setup and knock down. Fortunately, the setup in 1932 was less complicated than it is today.

To be successful, midway/carnival companies needed a good string or route of fairs. The operators often made contacts with fair directors at the annual conventions held for agri-

Conklin and Garrett Shows train, Western Canada, 1920s. The midway was a gilly show. This meant the show travelled exclusively by rail. At the zenith of the gilly-show era, Conklin and Garrett Shows travelled in a train of as many as sixty rail cars. At each stop, local teamsters were contracted to transport the baggage to the fairgrounds. Gilly shows died out when trucks became readily available. It was a lot of work to unpack, set up, take down and repack a whole midway from a train each week. Gosh, it's still a lot of work even in the era of the transport truck.

cultural fairs. At these conventions, the carnival/midway operator sold the fairs on the merits of his operation. A lot of salesmanship was required and sometimes the truth got bent a little.

Of course, the larger and more profitable fairs were highly sought after. Canadian fairs were unofficially divided into "A," "B" and "C" fairs, with "A" fairs being the largest. The top midway/carnival companies always aspired to a profitable string of "A" fairs. The fairs themselves would organize their dates so as not to compete with each other for midways, exhibitors or entertainment.

The Western Canada "A" fairs circuit was organized in 1910. It originally consisted of Calgary, Edmonton, Prince Albert, Saskatoon and Regina. Brandon later replaced Prince Albert. By banding together, these "A" exhibitions hoped to lure a major midway from the U.S. to play all five dates and strike a better deal in the process. This circuit quickly became one of the top fair circuits on the North American continent. Over the years, it has attracted all the big-name midway companies, including Johnny J. Jones, C.A. Wortham, Royal American Shows, and later, Conklin and Garrett. Following the success of this class "A" circuit, several class "B" circuits were established. Eventually all midway companies established a pre-planned circuit of fairs.

Midways thrived on the "law of density," which stated that the closer you packed the patrons together, the more they would spend. Midways have always organized themselves into tight spaces where the patrons are crowded close to the shows/rides/booths. A glance at any of the midway photographs in this book will prove this. If you see open spaces or lots of room on the midway, either the midway has not opened for business or they are having a really bad day. The midway also thrives on darkness. Electric lights and neon lights have made a big impact on the midway. There is a different feel to a midway at night. Sure, they run during daylight hours, but darkness adds a certain mystique and excitement to the show. Midways transform with the dark.

Carnival/midway companies have changed greatly over the last century. The earliest shows were short on riding devices and long on tented sideshows. Ferris wheels and merry-go-rounds were the only options before 1900. So, carnival companies relied on live shows to provide the entertainment.

Vaudeville-type stage shows, which included musicians, variety performers and comedians, were common. They usually involved dozens of employees including whole stage bands. The shows were expensive to produce due to the large number of performers needed. Smaller companies naturally had smaller entertainment contingents. In order to maximize use of these acts, the band often roved the grounds, playing between stage performances. As the old adage goes, music soothes the savage beast, and the midway bands often played during thunderstorms to calm excited animals. When announcing a show, barkers or spielers would line up the actors, musicians and the rest of the crew in front of the tent before each performance to help entice the paying public to enter the sideshow. This was called a bally. Decorative fronts were also used to sell the show. Some of these fronts were done by artists who took liberties with reality or stretched the truth about the inside acts, but all was fair on the midway. Eventually these vaudevillian/theatre acts died out. They were just too expensive. Why, the performers were drawing $18 a week each!

The inventory below was stored at a Columbus, Georgia, warehouse

Walsh & Wells Greater Shows Inventory List

2 X 68 ft baggage cars on lease from Hotchkiss-Blue RR Car Co. Lease and track storage paid through Feb. 1926

Item	Price
1 36 ft Hershell Merry-go-Round, band organ, LeRoi power unit, ticket cages	$2,300.00
1 No. 10 Eli (ferris wheel), Eli power unit, ticket box, crates, etc	2,200.00
1 Jones Merry Mix Up; 18 ft tower, LeRoi power unit	800.00
1 Kiddie Seaplane, 1 kiddie chairoplane, fence, etc.	325.00
1 4-octave Una Fon and batteries	145.00
1 10 X 10 office top with awning, floor poles, etc.	82.50
1 Oliver typewriter	25.00
1 army field desk, stock ticket box, 30 rolls misc tickets	38.77
900 ft (approx) slugging cable, tape, switch	90.00
1 Main Entrance Arch	62.85
1 bill trunk, apx 200 sheets pictorial posters 1's, 3's, 9's - not crosslined	16.52
1 solid canvas 14 X 24 pictorial Oriental Theatre, banner front	65.00
1 solid canvas 14 X 24 pictorial "AT" show banner front	65.00
1 30 X 40 M Kerr Co. top for Oriental theatre, poles, stage	182.00
1 30 X 40 Martin Tent Co. top for "AT" show, ring, poles	160.00
1 20 X 30 Baker Lockwood top for Buell walk-thru Flanders Fields peep boxes and extra set pictures, Barbary Coast	245.90
1 12 X 18 snake show top, banner front, inside pit, etc.	95.00
1 16 X 24 Plantation pictorial canvas front, bally, ticket boxes, 30 X 40 top, stage masking, plank seats	465.00
1 10 - 1 Beverly Bros. 20 X 60 top, string banner line, bally, boxes, inside pits and platforms, blow-off canopy	688.62
1 16 X 20 top, 3 banner front, Nelsons', Yolo & King Capakuli mummies	308.72
1 box tools, sledges, tapes, layout pins, etc.	41.00
1 Deitz hand drive, gasoline floss machine	46.00

for the winter of 1927–1928. It was never claimed the following spring, which likely signalled that the carnival company had gone bankrupt or the owners had met with a calamity in the off-season. The inventory list gives good insight into the composition of an early midway and the investment involved.

A typical midway of the early 1900s was the Boyd and Lindeman's Mighty Midway. In 1925, it played the Peterborough (Ontario) Exhibition. It was a typical gilly show, rolling into town on 30 railway coaches. It featured 7 rides, 15 shows and 52 stalls or booths manned by 375 employees. This horde overwhelmed the local hotels, and most of the workers were forced to camp on the fairgrounds. The featured mechanical rides were the Whip, the Seaplane, the merry-go-round, the Ferris wheel, Over the Jumps (an improved merry-go-round), the Merry-Mix-Up and the Caterpillar. The shows included the following:

Wild West Show
Law and Outlaw
[wax show]
T. W. Kelly's Circus Sideshow
[a freak show]
Congress of Fat Ladies
[featuring the Charleston Sisters]
Norman Shield's Animal Show
[40 "freaks of nature"]
Motordome [motorcycle show]
Midget Show
Ali Pasha's Arabian Nights
Congress of Athletes
World War One Exhibit

The fifty-two stalls included wheels of fortune, dart games, shooting galleries and other "diversions."[5]

Conklin and Garrett Shows at Weyburn Fair, Saskatchewan, 1920s. A huge crowd has gathered to watch something on the track, perhaps the human cannonball or a horse-race. In the centre is the sideshow labelled Physical Culture Stadium, a wholesome departure from the previous freak shows.

Midway lineups changed rapidly after World War II. In 1946, Patty Conklin won a five-year contract at the Canadian National Exhibition. He invested the princely sum of $250,000 to supply the following lineup:

RIDING DEVICES	KIDDYLAND	SHOWS	CONCESSIONS
1 Moon Rocket	1 Merry-go-Round	Water Show (modern and up to date)	2 Bingos
2 #16 Eli Ferris Wheels	1 Rotary Whip	Wild Animal Circus	10 "Flashers" (games) 20 ft. wide
1 Tilt-o-Whirl	1 Boat Ride	"Strange as it Seems" or Believe it or Not	1 Dining Hall
1 Spitfire	1 Ferris Wheel	Musical Girl Show	1 Candy Floss
1 Flying Scooter	1 Chair-o-Plane	"Miniature Monkey Land"	1 Candy Apple
2 Rolloplanes	1 Roto Scooter	Display of Canadian Small Animals	1 Popcorn
1 16-car Octopus	1 Aeroplane Ride	Penny Arcade	1 Refreshment Booth[6]
1 Sky Dive	1 Pony Track	"Glass House"	
1 Streamlined Caterpillar	1 Miniature Train	"Over the Falls" funhouse	
1 Hey Dey	1 Auto Ride		
1 Looper			
1 Auto Speedway			
1 Auto Scooter			
1 Merry-go-Round			

This 1946 lineup, when compared with the 1900 Bostwick-Frerari Carnival, offered a larger number of mechanical rides and a declining number of sideshows. Kiddyland had emerged from the mists of history to become a full-fledged sub-midway. The number of joints or stalls had declined, although independents were no doubt present in large numbers. Many of these rides, shows and concessions can still be seen on the midways of today, in more modern forms, of course.

The Rides

The oldest and most perennially popular mechanical ride is the merry-go-round or carousel. Originating in France, the merry-go-round arrived in North America via Britain in the 1870s. The earliest models contained seats shaped like sailing ships, balloons, gondolas, camels, lions, centaurs, wolves, donkeys and cows. Exotic creatures from other continents, such as tigers, giraffes, hippos and kangaroos, were added. The famous racehorses of the modern era became common only after 1900. The galloping, jumping horse

or the "up and down" merry-go-rounds gradually replaced the sedate, old-fashioned models. Gradually the handcrafted merry-go-round horse became a work of art in its own right. Craftsmen vied to produce elaborately decorated steeds. The craftsmanship reached new levels of artistic impression with each passing year. If a midway had to feature only one ride in its lineup, it was a merry-go-round. Amusement park carousels were much larger than their travelling merry-go-round cousins. Larger models could have as many as one hundred mounts arranged in rows up to five deep. But the demands of setup and knock-down limited the midway versions to a reasonable size. Even today, no fair seems complete without a merry-go-round.

Kiddyland rides, Edmonton Fair, 1927. The boat ride wouldn't look out of place in the midway today. Neither would the junior Ferris wheel in the background. However, the sideshow tents are long gone from present-day fairs. What sideshows do you show children anyway? The bug house?

Second only to the carousel among midway rides is the venerable Ferris wheel. Debuted at the Chicago World's Fair of 1893, the big wheel soon became a midway staple. The later versions were a little less imposing than the very first Ferris wheel, which stood 261 feet high (80 metres). Again, the logistics of moving every week kept the size of the big wheel in check. However, various versions of the Ferris wheel were invented to add a bit of spice and element of danger to the big wheel. These included double wheels (one above the other) and twin wheels (side by side). One of the big drawbacks of the Ferris wheel, no matter its type and besides the difficulty of setup, is the time required to load the patrons on the ride.

One of the big amusement rides that never caught on with the travelling midway companies was the roller coaster because it requires a huge, complicated and expensive base. This in itself doomed its appearance at the vast majority

of fairs. Only a few of the very large exhibitions such as the CNE and the PNE generated the gross dollars, length and attendance to make permanent coasters feasible. In each case the coasters were left standing to rest between fairs — sometimes for up to fifty weeks of the year.

When roller coasters first appeared in North America, they were considered the ultimate thrill ride. The first coaster, which opened in 1884, cost a nickel per ride. The owner recouped his investment in three weeks. The first looping coaster was considered so dangerous and thrilling, spectators were charged a dime just to watch others ride. But for most of the travelling midway companies, small, or kiddy, coasters were as far as they went in the roller coaster business.[7]

Gradually, over the years, a veritable plethora of riding devices have come and gone from the carnival/midway business. Some of the earliest (circa 1900) bore such romantic names as the Golden Chariot, Merry Mix-up, Whip, Over the Falls, and the Ride-eo. As technology advanced, the mechanical rides became more elaborate, often with names geared to the times, such as the space-age sounding Gravitron. Some of the newer rides bear an uncanny resemblance to earlier models. (Compare the Whip to the Tilt-a-Whirl of today!) Some rides, such as the Ferris wheel and the merry-go-round, have changed very little in a century. And other rides defy description. They must be seen to be believed.

The oldest and most reliable of midway rides: the merry-go-round, Edmonton Fair, circa 1920. This ride could not look out of place at a fair today, although the fashions certainly would. Merry-go-round or carousel animals became art forms and collectors' items. Whole catalogues have been devoted to their styles and designs. The merry-go-round was cheap, easy to set up and a crowd pleaser, especially with children. Every respectable midway had one. Some of the other rides in the background look a little more sinister and dangerous than the innocent merry-go-round.

Bumble Bee Ride, Edmonton Fair, 1920s. This kiddyride bears a remarkable resemblance to similar rides today. I don't think you will see many parents on kiddyrides today, and children certainly no longer wear ties!

The Shows

Fairs on the West Coast often had trouble getting good midways and stage shows. They were just too far from the epicentre of the amusement show business on the East Coast. One director of the Vancouver Exhibition complained that all they could book for the 1918 fair was a "second rate" Meyeroff Shows affair. Smaller fairs with smaller crowds had even more problems. With so many fairs crammed into so few weekends, it was sometimes impossible for small fairs to attract a midway of any description. As one newspaper reporter sarcastically observed, "Small fairs were often neglected by all but the most down-at-the-heel medicine vendors and lightning-rod agents."

The *Vancouver Daily News/Advertiser* described the first Vancouver Fair midway in 1910 as follows:

VANCOUVER'S FIRST EXHIBITION

HASTINGS PARK

AUGUST 15, 16, 17, 18, 19 and 20

Horse Show, Dog Show, Poultry Show, Industrial and Agricultural Exhibits.

Band Concerts Every Afternoon and Evening

Splendid Free Vaudeville Shows Afternoon and Evening in Front of the Grand Stand.

ON THE SKID ROAD WILL BE SEEN — Wrestling Competitions, Foolish House, Push House, Snake Charmers, Plantation Shows, Dancing Girls, and all the FUN OF THE FAIR.

Special excursion rates from all parts of British Columbia at less than single fare for the return journey. Boats will land passengers and exhibits at Exhibition wharf, one hundred yards from the gates. From Vancouver the grounds may be reached by C. P. R. special train, by boat to Exhibition wharf. City cars right to the gates at city fares.

On The Skid Road

Petrified Women, sacrificial crocodiles from the sacred river Ganges ... to say nothing of the numerous Salome dancers, Spanish Carmens, Dutch comedians and chorus girls, are some of the attractions that are being offered the visitors at the fair this week. Everything new and novel in the Amusement line, every means that the human mind can devise to gather in the spare nickel, dime or quarter of the Amusement seeker, is now in operation.

To do the Skid Road properly would take at least half a day. To witness each show, see all the dancing attractions ... ride the merry-go-rounds and be entertained at the half hundred other resorts, would require a sum of money not less than $5. And to eat a sandwich at each of the stands, drink lemonade at all booths or smoke cigars at each tobacconist would require at least another $5 bill.

Starting in at the south end of the road, the first concessionary is a rotund gentleman, who proclaims in stentorian tones the fact that he has in his show a petrified woman, recently unearthed in Arizona, and to prove the truth of this assertion, invites doctors and scientists to visit the attraction free. Proceeding North, the "educated horse" is the next show that strives to amuse the public. The promoters promised a number of things from this animal, calling him "Independence, the horse with the human brain." After passing several cigar and confectionary stands the visitor comes to the Spanish theater, a place where Oriental dances, a burlesque show and singing and dancing performances are given ... Down on the east side of the Road a knife rack, the Egyptian queen, a palmist, and ... the sacred crocodile of the Ganges, 200 years old as the "spieler" says, comes next in the line of amusements. Another dancing girl show, this one with genuine Dutch comedians, can be seen on the same street. Candy booths galore come next. Rounding the corner and turning up the South Road, the visitor is bewildered by the cries of the dozens of "barkers." On this Street the athletic arena, the maze of mystery, two dancing girl shows, and the Merry-go-Round are the principal attractions.[8]

Carnies at a "Front End" gaming concession, 1920s. The actual games are on the table in front. The carnies are well dressed and the whole booth seems quite presentable to the public. Prizes to be won grace the background.

The midway just prior to opening. Shows with their fancy fronts line both sides. In the rides row are the Frolic, the Underground and the Ferris wheel. An ice-cream/soft drink stand is setting up in the foreground. Note the tarpaulin roofs have not yet been added. The white picket fences give the midway a homey, backyard feeling.

In Vancouver the midway was called "Skid Road," a term rarely used in other parts of the continent. And dunk-tanks were a new and exciting booth in 1910.

"Freak" shows were a big part of the travelling midway/carnival in the first half of the twentieth century. These generally consisted of people or animals with unusual physical characteristics. They included fat people, dwarfs, half men–half women, two-headed creatures, Siamese twins and just about anything the mind could imagine and a lot it could not! Liberal use of imaginative advertising accompanied freak shows. The most common advertising technique was the use of painted canvas "fronts." These elaborate signs were painted by special artists who certainly had vivid imaginations, or who never saw the actual "creatures." The larger midway/carnival companies had their own sign or front painter on permanent staff. It was necessary to produce dozens of different fronts every year as acts changed or as the old signs were destroyed. One such front for the "Penguin Boy" showed a youth with "perfectly proportioned wings perched atop an iceberg. Inside [the tent], what the audience found was a short man with stubby arms smoking a cigarette."[9]

Finding candidates for these shows was a difficult and time-consuming business. The promoters scoured the world looking for oddities and people with deformities. One promoter grumbled that World War II had really damaged his business, since many of

Freak show, 1930s. This was a "10 in 1" show: ten freaks for the price of one admission. The hand-painted fronts most certainly exaggerated the actual show; but then, imagination was a strong selling point on any midway.

72

The Finnish Giant. Clearly he survived the war as this 1950s photo from the Edmonton Fair proves. Notice how the hat and clothes enhance his size? All's fair on the midway.

his top acts came from Europe. He had his eye on a Finnish giant who towered 8 feet, 4 inches tall, ideal material for the "world's tallest man" show. To the promoter's great distress, this behemoth was trapped in the Finnish army where, it was worried, his height made him an easy target. If he survived the war, the tall Finn was expected to gross $300 a week on the fair circuit and another $100 posing for photos. This same sideshow promoter foresaw an even more dangerous threat to his vocation than a mere world war. He complained in an interview, "This country isn't producing freaks so much anymore. These sanitary codes and health laws have a lot to do with it. Improve the public health and they beget fewer monsters. You get sanitary codes and you don't get many three-legged and two-bodied people."[10] I bet you didn't realize the health laws had such a downside? This wasn't the only bad experience the same promoter had with modern medicine. One of his acts was a little person named Col. Tiny. One day the Colonel was hospitalized for a severe illness. The doctors had found a pituitary gland problem and treated the performer for his ailment. This caused Col. Tiny to grow. Grumbled the promoter, "He didn't grow big enough to be a man, but he grows big enough to where he isn't a very good midget. You would think a doctor, a fellow with a college education, would have more sense."[11] One wonders how long Mr. Sensitivity would last in the modern world?

Even the freak shows had a pay scale, a difficult concept to comprehend for people outside the business.

The heavyweight entertainers, CNE, 1930s. Sideshows were still obsessed with the unusual. Besides being on display, these performers sang and danced. I'm not sure who the bloomers hanging over the doorway belong to, but they are a good example of overimagination.

In the 1930s, average salaries for performers ranged from $25 to $125 per week. At the low end of the scale were the human skeletons, also called cigarette fiends. Fat men earned $35, while fat ladies grossed $50 per week. Bearded ladies commanded up to $75 (depending on the hairiness) and giants pulled down $80. The top salary of $125 was garnered by the "baby with four legs."[12] Salaries were clearly based on the supply of performers, with the really rare commanding top dollar. Nevertheless, being a freak was not a high-paying profession. When the carnival companies disbanded for the winter, many were forced to seek temporary employment elsewhere. The better-off went south for the winter, gathering together in an informal community at Gibsonton, Florida. By the second half of the twentieth century, these shows began to disappear on the fair circuit. A better-educated, more sensitive and more informed public became increasingly skeptical of these shows. Public opinion turned against people and animals with deformities exhibited in such a manner. One of the last freak show promoters bewailed the changing attitudes in a 1958 interview: "You cant get good freaks anymore. Seems like they're all dying off. They take em and put em in institutions now. They don't want em exposed. Now I ain't going to mention any names, but I know an insane asylum where there's three good pinheads right now. But you can't get them out."[12] Oh, the trials and tribulations of a changing world! Thankfully, these show have virtually disappeared from fairs.

Monkey auto racers, Conklin and Garrett Shows, 1920s. This show combined the animal world with the machine age. Collisions are featured in one panel, perhaps a forerunner of demolition derbies. Even in the 1920s, the thrill of the crash was a big drawing card. Today, the monkeys have been replaced by humans in demolition derbies.

Animal acts were always popular shows, especially exotic creatures from other continents. Performances by specially trained animals such as dogs, ponies and monkeys were a part of every midway. The larger or more unusual the creature, the better. In a world where few people got the opportunity to travel or even read about far-off lands, exhibits of people/life/animals from other continents were popular. Historical accuracy was not always a strong point in the sideshow business. One 1909 act was billed as "Fatima, a Bedouin from Nineveh, performing the mystic anaconda dance, exactly as danced by Hypatia in Holy Writ." This billing was a hopeless hodgepodge of historical inaccuracy. Bedouin Arabs are found in North Africa, while Nineveh was an ancient Assyrian city separated by hundreds of miles and a thousand years from the Bedouins. Anaconda snakes are found only in the jungles of South America. Who Hypatia and

Inside the water show. The bathing belles and crew pose for the photographer. Besides the swimmers, the show involved a brass band, several clowns and numerous speilers, ticket sellers, security staff and promoters. It took a lot of personnel to put on a show like this one.

the Holy Writ were has been lost in the sands of time. But as one writer put it, "Fairgoers were interested in anatomy, not history or geography."[13]

Even different peoples from North America were presented: cowboy and Indian shows in Eastern Canada, and lumberjack shows on the Prairies, to name a few. With so many wild animals around, accidents and catastrophes were not unheard of. Two lions escaped their pens at the 1909 Calgary Fair and ate a trained pony. A trained mastiff and some carnival workers cornered the lions and several cowboys lassoed them. Another tragedy occurred when the wolf dogs got loose and annihilated the rat show. They were credited with devouring 500 rats at a sitting.[14] It was likely that many of the victims were not eaten, but used the confusion to escape to greener pastures. One act touring Manitoba was called "The Wild Woman from Lowlands of Madagascar, surrounded by a thousand crawling, hissing reptiles from every clime; eats, breathes and sleeps where a dog would not live an hour." Two police constables debated charging the show with either fraud or slavery![15]

Another popular entertainment was the water show. This could involve anything from high divers

Water show, CNE 1950s. Water shows still emphasized beautiful girls in a mermaid-style theme. Women in bathing suits really turned heads.

75

to synchronized swimmers to "mermaids," performing in a water tank or pool, often enclosed in a tent. Young women were the preferred performers. After all, to the targeted male segment of the audience they did look good in bathing suits. The water shows must have endured some cold water days, there being no heated pools. Diving horses were also a frequent novelty act. Many farmers were mystified that the noble horse could turn into an aquatic champion. Motorcycle-stunt acts were very popular. The riders did tricks and stunts in a portable velodrome. To add some spice and a sense of danger, animals were sometimes thrown into the mix. Monkeys seem to be the common addition, but one act used lions. A newspaper reporter noted in 1939 that "Marjorie Kemp's Bowl of Death" motorcycle show had acquired a new lion. The old one had retired to a zoo.[16] The new lion seemed to be a rather unhappy passenger, adding an element of uncertainty to the show.

The earliest midway/carnival companies introduced many rural patrons to their first motion-picture show. Portable movie theatres toured with the midway company, and for a dime the rural resident could watch a short film or newsreel. Of course they were silent flicks, but the novelty of motion pictures attracted huge crowds. The 1900 Saint John (New Brunswick) Exhibition advertised the "Wonderful War Graph: Moving Pictures of the Boer War." A lineup in a Northern Ontario midway in 1905 included: "The Great Train Robbery," "Miniature World" and the "Great Toronto Fire." Full-length movies were rare and not popular with midway operators: they took the patron out of circulation for too long at a stretch. By the 1920s, movie theatres were more common in Canada even in rural areas, and the novelty began to wear off on the midway.

Midway/carnival operators were always trying to keep up with current trends. One way to do this was to build acts on national or world events. Obviously the two world wars were popular topics. A little patriotic fervour, with a dash of profit, made for a successful show. Natural disasters such as floods, volcanic eruptions and earthquakes all reeled in the fair-goer. Famous crimes and murders also attracted attention. A popular display in 1935 was Bonnie and Clyde's "last getaway car." One entrepreneur stole the shot-up car from a police pound and took it on the fair circuit. It caused quite a stir until law enforcement officials seized the vehicle. To keep their momentum, the midway operators bought a similar car, shot it up, substituted the new vehicle and proclaimed it the original getaway car.[16] Who was to prove them wrong?

To many fair-goers, midways had a dark side: gambling and games of chance. Gambling and games of chance had been part of fairs as far back as medieval England. Many agricultural purists lobbied long and hard to have midways/carnivals banned or curtailed because of the gambling. Carnivals and midways received a bad reputation because of it, and to a certain extent this stigma still hangs over the midway today. Games of chance took many forms: roulette wheels, ball throwing, dart throwing and shooting galleries were just a few. The element of the midway who practiced gambling or games of chance were labelled "sharpers," "conmen" or "fakirs." The word fakirs comes from Arabic meaning beggar and refers to a religious mendicant. How this term was transferred to a carnival conman defies explanation.

In the early 1900s, vigorous steps were taken to clean up games of chance at fairs. Ontario appointed four constables to study the problem and help clean up the midways. A 1903 provincial statute proclaimed:

"The officers of the society shall prevent all immoral, or indecent shows and all kinds of gambling and games of chance, including wheels of fortune, dice games, pool tables, draws, lotteries or illegal games at the place of holding the exhibition or fair, or within 300 yds thereof, and any association or society permitting the same shall forfeit claim to any legislative grants during the year ensuing."[17]

Forfeiting their government grants! Now, that hits fairs right where it hurts!

Subsequently, three Ontario police constables spent the fall of 1904 hard at work rooting out fakirs and sharpers at fairs. They reported at the 1905 Ontario Fairs Convention:

"The gamblers and sharpers who attended the exhibitions find their work so profitable many of them make it a regular business. Their routes are mapped out long in advance. Every possible precaution is taken to guard against arrest. On reaching the place where the exhibition is to be held, the

Inflation takes its toll on the ageless fish pond game. In the 1960s it cost 10¢, and by the early 1970s, 25¢ to "win a prize every time." By 2000, the cost had inflated to $2 a chance, or three for $5. Regardless of price, the game never changes.

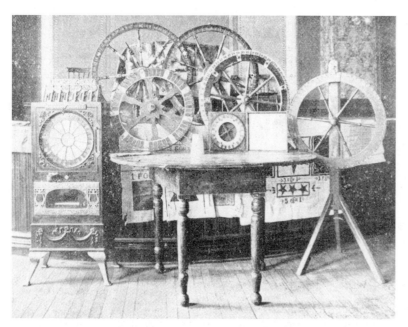

A sample of heinous gambling devices seized at 1904 Ontario fairs. Such tools of the devil were considered the root of all evil at fairs a century ago. Today, they don't look so threatening. Times change.

local police are first seen and, if necessary, bribed. Money is then offered to the officials of the fair for the privilege of operating at the exhibition [the officials of one fair tried to put the provincial detectives off the grounds for interfering with the gamblers]. If a visit from a Provincial detective is feared, the gamblers appoint a man or men to watch the arrival of trains and the hotels for the appearance of such officials. At the first sign of danger the signal is given and the games of chance disappear as if by magic. So well did the regular provincial detectives become known two years ago it was deemed advisable this fall to employ two new men, especially for this work. Before the end of the season these new men were so closely watched it became difficult for them to make arrests. At a fair in the Niagara district a gambler offered one of the detectives $50 to leave the grounds. This will give some idea of the profits this man must have expected to make. Eighteen sharpers at one fair in Southwestern Ontario pitched in and raised $100 as a bribe for Detective —? [who had been spotted in spite of his best endeavors] to leave the town.

"At the fairs I visited I found a great many sharpers, fakirs and gamblers, some of whom were very clever. They seem to have plenty of money and do desire no other occupation during September and October, as they claimed they could make enough off the farmers in that time to keep them all winter. I saw as many as 18 sharpers at one small B fair, with an attendance of about 3,000 among them being three whom I had arrested at a previous show. Their fines were paid in each case out of a large roll of bills they had in their pockets.

"I did not find a crook who did not know provincial detectives Rodgers, Greer and Murray at a distance. They made this a study. The fakirs seem to lay out their sections as economically as possible. In some cases I saw the same people at different fairs for four or five days in succession. In such cases a change of detectives is required to catch them.

"I believe that in the counties of Kent, Essex, Elgin and Middlesex, I saw up to a hundred different people making their living by fraud at agricultural exhibitions. It seems to be the desire of all the officials at the fairs to have a straight, clean show, but they are easily deceived by the sharpers, who pay so much to one spotter who occupies an elevated position and when necessary gives the alarm to his confederates.

"It is time these characters were driven from our exhibitions, and the latter made what they were intended to be, a benefit to agriculture. The majority of directors would like to see their fairs free from these features. Other directors boldly state that the public likes to be fooled, and that their exhibitions need the money the gamblers are willing to pay. The worst feature is that the public is not only fooled, but robbed by many of these men.

"At an exhibition in the Niagara district last fall a gambler had an outfit, worth about $75, fitted up specially for this purpose. It consisted of the table underneath which, out of sight, was a drawer containing 23 dynamos. He had loaded dice, and invited the public to try their luck betting to see who could make the highest throws. By simply leaning against the edge of the table he could control an electric current and cause the loaded dice to fall any way he desired. The method adopted was to allow the person throwing to win several times, and at the same time gradually increase the size of the bets. As soon as a sharper was sure his victim was betting all his money he immediately cleans him out. The department has received reports that men have lost $150 and $200 in an afternoon through swindles of this nature. The gambling outfit, just mentioned, was confiscated, and will be on exhibition at the February meeting of the Fairs and Exhibitions Association in Toronto that the directors of the societies who attend may have a chance to try their luck."[18]

To be sure, some carnival/midways were poor ambassadors for the industry. In 1951, Medicine Hat Fair contracted a midway sight unseen. The operators showed up on fair week with five broken-down rides and twenty ladies billed as "fortune tellers and palm readers." Realizing their mistake, the directors made an attempt to rid the fair of the so-called midway and replace it with a more reputable firm. Another midway was promptly procured, but the original outfit refused to leave. Consequently, the next day a posse of mounted cowboys thundered into the fairgrounds and trashed the "gypsy camp" in a scene right out of an Old West movie. The so-called gypsies were gone by nightfall. Problem solved![19]

Conmen had many tricks to fool the unsuspecting "country rube." One such scam was called the

A travelling medicine show, Saskatchewan, early 1900s. Kickapoo brand products were obviously popular at the time, but this brand name seems now forgotten. "Do it yourself" or "on the spot dentistry" is also clearly advertised. A rather gruesome collection of teeth in sealer jars is either meant to testify to the skill of these "dentists" or draw attention to the display. Somehow, these gentlemen don't inspire much confidence.

The Roller-Boller combines rollercoaster and water. Obviously this ride is the ancestor of the modern water ride. It was elaborate and costly and must have operated on more occasions than just the Canadian National Exhibition.

Kleansol soap gimmick. A salesman would extol the cleaning virtues of this miracle soap. To demonstrate, the conman would take a clean, white handkerchief from his pocket, wipe the grease off a handy wagon axle and deposit the soiled hankie in a tub of cold water. A small cube of the miracle soap was selected and a few flakes shaved into the tub. In seconds the tub was engulfed in soap suds. Then the handkerchief was extracted from the foam white as snow. The suitably impressed onlookers had no way of knowing that the axle was greased with tar soap or fake grease that dissolved instantly in water. The con artist proceeded to sell his miracle cubes at 25¢ each, a seeming bargain. But Kleansol soap retailed at 5¢ a cake. The fakirs cut each cake into a dozen miracle cubes.[20] All for a good day's profit!

The sharper who auctioned off items — well, this was part of another con commonly found at the fair. Fake or cheap jewellery was the favoured commodity. The sharper started by maintaining he was just advertising for a larger company and had a few "samples" to give away. Everyone in the crowd was given a small gift, say, buttons. A gift attracts a throng of people, including many children. To dispose of the children and the less serious, the sharper said he would pay a dime for the buttons. The dime was given to the onlookers but the sharper would also return the buttons as a symbol of the sharper's generosity. Thus impressed, and intrigued, the onlookers waited for the next gift, maybe a pair of cufflinks for a quarter. Once again the sharper would up the ante by giving back both the quarter and the cufflinks as a gift. With the crowd suitably impressed, the sharper moved to the third and final act. He would offer a ten-dollar watch chain for only a dollar. From the crowd, an accomplice of the sharper would openly and loudly challenge the deal. The sharper would then display his prowess by tongue-lashing the lackey into quiet submission. Duly cowed, the audience would meekly hand over their dollars, but this time receive only the watch chain. The value of all three gifts might total 15¢, leaving the sharper with a tidy profit. Few people complained, unsure of the value of their gifts, and unwilling to endure the famous tongue-lashing, they had just witnessed before them.[21] Such scenes were commonplace at many early-twentieth-century fairs.

While the agricultural purists, fair reformers and many other groups railed against the conmen and fakirs on the midway, to others the games of chance were just great. For many rural residents, the flash and dash of the midway was an exciting break once a year from the tedium of farm life. In the age before both television and travel shrank the world, the midway, and indeed the whole fair, was a wonderful world of the exotic, the new and the unusual.

The charms of the midway were especially strong for children. Many readers may pause and remember their childhood experiences on the midway. Sadie Marsh of Granum, Alberta, related her midway experience as an eight-year-old farm girl. She amassed the princely sum of 25¢, earned by picking potato bugs for her father.

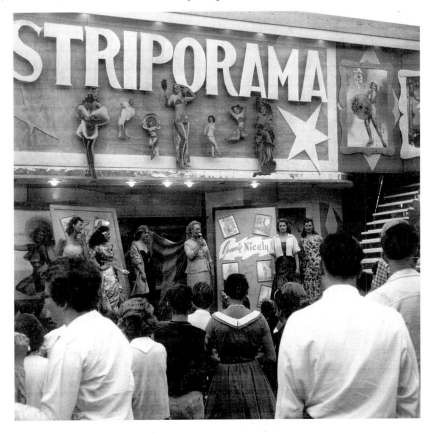

After years of cleverly disguising strip shows under such names as "burlesque," "hootchy-kootchy," "exotic dancers," etc., this 1950 sideshow gets right to the point. The enraptured audience listening to this spieler (a woman at that!) contains a goodly portion of women.

"Fair day finally arrived and my sister Myrtle and I were dressed in our white eyelet embroidered dresses, with our hair in ringlets, after sleeping with our hair done up in rags. As we drove to the grounds the first thing I spotted was a booth full of dolls with a wheel of fortune: 25¢ a chance. I was out of that buggy like a shot heading for those dolls with my mother yelling, 'Sadie, don't you dare spend all your money on that game of chance.' I just kept going and plunked my hard earned money on a number.

Conklin's Midway at Yorktown Fair, Saskatchewan, 1937. Yorktown was a B fair, but still attracted good crowds, as this photo shows. This is a typical midway setup with rides in the centre, sideshows down one side and games and concessions completing the rest of the circle. It also emphasizes the importance of the canvas tent to the fair industry.

That wheel spun around and stopped on my number. I won and picked out the biggest doll in the booth! I climbed back in the buggy and spent the remainder of the day holding my doll with my Mother lecturing me on the pitfalls of gambling and games of chance." [22]

To the bored Prairie girl, the midway must have seemed like paradise!

Games of chance could also be easily rigged. Cans to be knocked over could be weighted or set off-parallel. Holes for balls to pass through were either too small or just barely large enough. Basketballs could be overinflated or the hoops undersized. Dart games had the balloons underinflated, making them harder to break. "Toss games" had prizes hanging so low from the ceiling, the thrower could not get the proper trajectory to land his coins safely in the dish. Crown and anchor wheels were equipped with brakes. The corks in air guns were shoved in too far, thus limiting their velocity. [23]

Another gimmick involved a math game. A player would knock over eight sticks with a ball. If the sum of the numbers totalled a figure listed in red or green on a board, the player won. If the sum of the numbers was in black, the game owner won. Apparently the odds were two to one in the player's favour. Actually, there were as many black numbers as red and green combined. But the real con was in the addition. The quick-calculating sharper added so fast, the player could not keep up. A little mistake in addition turned a victory for the player into a victory for the sharper. Only math-whizzes could keep up. [24] All these tricks of the trade have been used on midways, yet they never seem to have effected the midway's appeal or attraction.

Other marks on the negative side of the midway ledger were the girlie shows. They sprang up after the tremendous success of "Little Egypt" at the Chicago World's Fair. Variously labelled girlie shows, hootchy-kootchy, cooch, burlesque or just plain stripper shows, appeared on many midways in the early 1900s. They were obviously aimed at the male segment of the population. To the farm boy, belly dancers and burlesque reviews represented a new and exotic part of life. One old burlesque joke went: "When I was a boy, my father told me I shouldn't go to the cooch show. Of course I asked, 'Why not?' And he said it was because I might see something I shouldn't see. Well, I went and he was right. I saw my father." The sideshow promoters found out early that sex, or at least the illusion of sex, sells. Very few of the so-called exotic dancers actually "took it all off." One longtime fair manager described a cooch show this way: "The girls would come out on the bally as a sort of preview. They were all pretty good looking and pretty well built, and they led you to believe that hot stuff awaited during the real show. You thought it got better inside, but it didn't."[25] But, oh boy, those rumours sure did attract attention! A typical example of fact versus fiction revolved around the 1930s show involving exotic dancer Jade Rhedora. Rumours circulated that she danced stark naked. One disgusted politician, after hearing numerous complaints, decided to check it out himself. The bally promised "complete nudity," and upon viewing the performance the outraged politician confirmed the claim. "Not so much as a piece of corn plaster interposed itself between the goggle-eyed public and the girl's epidermis," fulminated the infuriated legislator. However, the good gentleman was only another fooled patron. Rhedora was not naked, but rather ensconsed in flesh-coloured tights in a dimly lit tent. The ensuing publicity was the best advertising possible, and the show finished its season to sold-out audiences.[26] Even as late as the 1940s and 1950s, major midway companies still featured at least one cooch or girl revue. Such famous dancers as Gypsy Rose Lee, Bonnie Baker and Sally Rand were huge attractions for the midway companies. Sally Rand, gushingly referred to as the "fair enchantress," regularly drew crowds of 10,000 at the larger fairs. These high-class girl revues included the highest-paid performers on the midway and, from a purely business point of view, were worth every penny.[27] Many of these girlie shows were divided into segments. The customers were enticed by a spieler as the girls stood at the front of the tent. Inside was a room where the show was held. But sometimes the real show happened in a back room of the tent. Invitees paid another fee to witness the final act. This part of the show was for men only. As one spieler said: "No children, ladies or Sunday-school teachers are allowed!"[28] But even as the cooch shows reached their zenith, their demise was already underway. One carnie observed that "women were beginning to wear clothes that were tight, skimpy or revealing in public as part of their normal attire,

Zacchini the Human Cannonball draws an expectant throng. The net is not visible, but there is likely one somewhere amid the crowd. However, if the great Zacchini goes a little off target, he will surely be crowd surfing. In the background is the huge human "freak" exhibit. The front proudly proclaims "all freaks alive!," which may or may not have drawn much comfort for the performers. Also in this picture is a version of the swings and a coaster-style ride, still common in midways today. The swings were brought to you by Imperial Petroleum Products; sign sponsorship was alive and well in the 1920s.

and men could watch them all day for free."[29]

Gradually the carnival/midway companies began to clean up their acts. Gooding Shows set the example. In the early 1900s, they ran what was called the "Sunday-School Midway." Needless to say, they were in heavy demand among fairs. Other midway companies began to get the message. The games of chance were cleaned up. The girlie and freak shows disappeared and were replaced with an ever-expanding inventory of mechanical rides. Such changes occurred for a number of reasons. First fairs began to demand more rides at their exhibitions as they faced increased competition from amusement and theme parks. Television also played a role, bringing a withering array of shows, spectacles and entertainment once available only at the fair. Costs for the large stage shows, like the live bands, vaudeville acts and water shows, also escalated to the point where they were unprofitable. And as tastes, sensibilities and sensitivities changed, so too did the fair.

A Tough Business

The midway/carnival business has always suffered a dizzying array of ups and downs. Natural disasters, bad weather, world events and public whims have played havoc with the midway/carnival business. The biggest threat was the dreaded windstorm, called in carnie tongue a "blow down." High winds, tornadoes and storms occasionally destroyed the often flimsy midway setups. Rain and bad weather wreaked havoc with the crowds and caused huge financial losses. On one Saturday at the Calgary Stampede in 1942, weather limited Conklin's Midway gross to $9.28![30] The Great Depression too caused a shrinkage in fair-going and spending. In 1933, Patty Conklin was refused a $500 loan by a bank to start his fair season. Flat broke, his midway rolled into Timmins, Ontario, for a still date. After a rough start, the heavens smiled on Patty as a young miner who had recently struck it rich came through the gates and quickly spent over $3,000, thereby saving the show from bankruptcy![31] Wars, of course, took a terrible toll on the midway/carnival business. Many larger fairs shut down during both world wars, leaving a huge void in the midway circuits. When the CNE was closed for six years (1941 to 1946), Conklin and Garrett was without the crown jewel in its circuit. And just before this, terrible polio outbreaks caused attendance drops in 1937 and 1938. As the slogan goes: "Sometimes you win, sometimes you lose, and sometimes it rains!"

Bankruptcy was also a common cause in the demise of many midways. It was an uncertain business at best. The saga of the "World of Mirth" midway is typical of many in the business. Founded in 1930 by the merger of several smaller units, World of Mirth quickly became a force on the north-eastern fair circuit in Canada and the U.S. One of its prize stops was the Central Canada Exhibition in Ottawa. The Great Depression did not damage its financial health and the midway survived a five-year layoff for World War II. But in the late 1950s, the World of Mirth began to decline. From fourteen fair dates in 1954 it dropped to only eight contracts in 1963 with Ottawa still its number one spot. Accidents in consecutive years led to its final demise. In 1962, a ten-year-old girl and a four-year-old boy both fell ten feet to the ground from a ride called the "Meteor." The boy ventured outside the car, and the girl went after him. The girl died from her injuries. The very next year, a car

Midway scene with freak shows. The show at right has an extensive tent which indicates that it is a large sideshow. The painted fronts conceal the less attractive tent.

from the "Kilimanjaro" ride left its track and crashed into the ride's control booth. Nine people were injured. The back-to-back catastrophes led the Ottawa Exhibition to cancel its long-standing contract with World of Mirth Amusements. The loss of its key date caused the midway company to instantly fold and sell its assets.[32]

Although accidents were rare on midways and liability often does not rest with the midway companies, ride safety became a big issue. Any accident, no matter how small, was magnified by the press. Government inspections occurred regularly and new standards of safety were introduced. Some smaller midway/carnival companies were forced out of business. All carnival companies placed new emphasis on ride safety. And in an age when liability was becoming a big factor at all events and in all business operations, both midways and fairs were forced to take even more stringent precautions. Fairs in other words, were mirroring their times.

As in the past, midways today have been forced to adapt to changing times. There is more competition for the entertainment dollar. Theme parks can offer more elaborate rides on a year-round basis. Since such parks have permanent sites, they can boast larger, more spectacular rides; they do not have to knock down their operations and move every week. Television, movies, computers and video games have stolen some of the impact and aura of the carnival. The investment required by a midway/carnival company, both in technology and staff, has made many of the smaller fairs less attractive and financially viable than larger draws. Midways need a certain draw of people to cover expenses. This does not augur well for smaller fairs. After all, a fair without a midway is simply missing something.

But midways/carnivals have faced big challenges in the past, and they have become very adaptable in the process. When sideshows and variety acts died out, the midway/carnival operators quickly replaced the void with mechanical rides. One cynic I know now believes midways are dying; he maintains that midways now appeal only to young children and have lost their teen and adult patrons to other branches of the entertainment industry such as theme parks and video games. But the midways will surely survive and adapt because their strengths are flexibility and adaptability. The more things change, the more they stay the same.

WORLD FAIRS

The world fair idea is an old one. The medieval trade fairs were successful because they drew vendors and merchants from faraway countries. This international flavour was what made a trade fair great. In the 1700s, various nations in Europe began to play with the idea of holding a huge fair "open to the world." The basic premise behind such an exhibition was twofold: it served as a melting-pot trade and industrial exhibition where new ideas and advances in technology were displayed. It was also an opportunity for the host country to "show off." No nation ever sponsored a world's fair unless it wanted to show off its industry, technology and culture. World fairs were not limited to agriculture or even industrial technology. They were a great conglomeration of every aspect of human life. Many were called "industrial exhibitions," but they were more than just a forum to exhibit new technology.

The first world fair of the modern age was held in Paris in 1798. It was called an industrial exhibition, but thanks to a little disturbance called the Napoleonic Wars, attendance was limited. The world fair really struck success at the 1851 Crystal Palace Exhibition in London, England. The 1851 exhibition was named after its outstanding hall, the Crystal Palace, so named because it had glass panes for its walls and roof. The exhibition was designed to show off the industry and commerce of the British Empire, then at the zenith of its power and prestige. But exhibitors from all over the world were also invited, and it was the international flavour that made this fair so famous. Thanks to steamships and railways, travel for fair-goers was made possible from most anywhere in the world. The exhibition lasted five months and attracted 6 million visitors! So successful was this exhibition that other countries around the world immediately began to plan their own.

World fairs have continued from 1851 to the present day. Some have been successful, some not so successful. Most have lost money, but then profit is not the primary goal behind these shows. Education, learning and showing off were the big factors in 1851 and they still are today.

Canada staged its first world fair in 1967 to celebrate our centennial: a century as a nation was as good an excuse as any to host the world. Expo '67 in Montreal was a marvellous success with over 50 million visitors passing through the gates. It was widely praised for its architectural displays, site organization, technology and cultural presentations. International visitors were left with good impressions of Canada. Above all, it left Canadians feeling good about themselves. And hey, if Canadians want to show off, what better place than a fair to do it! The most recent Canadian world fair was held in Vancouver in 1986. The theme "World in Motion/World in Touch" emphasized advances in transportation and communication.

CHICAGO WORLD'S FAIR OF 1893

Holding a world's fair was considered a mark of prestige for any country. The United States announced it was holding a world's fair in 1892 to honour the 400th anniversary of the arrival of Christopher Columbus in the New World. Since many felt that an excuse was needed to hold a world's fair, this anniversary was considered as worthy an excuse as any. The city of Chicago was then selected to hold the fair. Unfortunately, preparations got behind schedule and the date was pushed back one year.

This famous exposition introduced hamburgers, carbonated soft drinks and postcards to the world. Such popular products as Juicy Fruit gum, Shredded Wheat and Aunt Jemima Syrup were also displayed for the first time. The first hybrid field corn (Reid's Yellow

Dent) was displayed to the public. Scott Joplin entertained with that new-fangled "ragtime" music. Western Union introduced a generation to Standard Time, which was conceived by a Canadian: Sanford Fleming. Electricity, the dynamo and Westinghouse's new alternating current (A-C) generator were all on display. The extensive use of electricity allowed for after-dark operations, a big advance for the fair industry — and the world.

This watershed exhibition allowed the newest technologies and inventions to be viewed by millions of Americans. Attendance surpassed the 25 million mark over the course of the fair. It was estimated that one in every fifteen Americans attended the great fair. And these fair-goers had the opportunity to see the newest trends, inventions and technology. It was an advertiser's dream! L. Frank Baum was so impressed during his

The very first Ferris wheel, Chicago World's Fair, 1893.

visit that, when he later wrote *The Wizard of Oz*, the Emerald City was directly copied from the "White City" he saw at the Columbian Exhibition.

One important legacy of the Columbian Exposition of 1893 was the appearance for the first time of an organized midway at a North American fair. Previously, carnivals and sideshows had operated independently and randomly at various fairs and exhibitions. At Chicago, the independent carnival/sideshow operators were brought together in one section of the fairgrounds: "the Midway Plaisance." This midway was a park mid-way in the fairgrounds, and "plaisance" was a French term for pleasure grounds. Into this section of the fair were jammed animal shows, freak shows, sideshows and almost every act imaginable. Carnival barkers or spielers were banned outside the shows, so some adaptable carnies did their spiel in pantomime. Another section of the midway was laid out as streets from exotic countries around the world. The most famous — and infamous — was "the Streets of Cairo" display, which gained notoriety for the thin womanly veils found on the byways in "Little Egypt."

The Chicago World's Fair had a slow start, and the midway operators were suffering from poor attendance, when a general meeting was held and the problem of how to attract patrons was discussed at length. One clever carnie, recognizing the power of sex appeal, suggested using a belly dancer named Little Egypt to draw attention. Little Egypt naturally drew negative press and hence huge crowds of curious fair-goers. The more the press condemned the act, the more patrons flocked to see what the hullabaloo was all about. Much of the belly dancer's reputation was carefully enhanced by midway operators, and enhanced with a lot of fantasy. One favourite ploy was to leave a transparent dress on display outside the show, implying, well ... Many imitators followed Little Egypt's successful act. One, "Fatima of the Thousand Veils," was actually a father of five! So successful was Little Egypt that exotic dancing, hootchy-kootchy and strip shows of various kinds became standard fare at most midways for many decades.

Little Egypt went a long way toward setting the reputation of midways, but what really sealed the success of the Chicago World's Fair was the introduction of the Ferris wheel. The idea of a vertical riding wheel had been in the plans for years, but it took a Pittsburgh engineer named William Ferris to actually make it work. An earlier contractor had backed out of the project at the last moment and Ferris had stepped into his place. His massive wheel stood 250 feet high and carried 1,440 people at one time! The Ferris wheel, combined with the allure of Little Egypt, saved the Columbian Exposition and started the midway/carnival business on the road to American legend.

After the fair, midway operators talked about joining forces to create super carnivals. These travelling shows were to contain a mixture of animal acts, games, sideshows, entertainments, and riding devices. The first travelling midway companies failed, but showmen still remembered the successes won at the Chicago World's Fair and persisted in their attempts. By 1900, successful midway/carnival operators were numerous in the U.S. and, naturally, they spread to Canada. And it all started with Little Egypt and the Ferris wheel at the 1893 Coluombian Exposition.[1]

Entertainment: See It at the Fair

THROUGHOUT THE HISTORY OF FAIRS IN CANADA, organizers have debated the issue of entertainment at agricultural fairs. To the agricultural purist, entertainment has little or no place at a fair. Fairs are simply meant to promote agriculture and rural lifestyles. But the fair without entertainment is … almost unheard of! The law of reality states entertainment is essential to the success of every fair; it is a great way to attract patrons.

The problem really lies in the degree of entertainment, or the balance between education and entertainment. This is not a new debate. It was an issue 200 years ago and it remains an issue today. It is a chicken-and-egg type of debate. To promote agriculture, you need people to come to the fair. To get people to come to the fair, you need entertainment. Each one needs the other. After all, fairs are supposed to be community festivals as well as educational events.

Horse Racing

Even the earliest Canadian fairs had an entertainment component. The original Hants County Fair held in Windsor, Nova Scotia, in 1765 made an allowance for a sports program. Midways and carnivals were unknown and sideshows were rare at early fairs. But sports—now *that* was something easily organized and appreciated. Competitions included human athletics, or the ever popular horse racing, no doubt the most well-attended sport at fairs. After all, horses were farm animals. Most farmers had one. And the whole show would be livened up with a little betting, either official or unofficial. In many communities the racetrack and the fairgrounds were the same place and facilities were shared. Indeed, almost every fairgrounds in Canada had a racetrack at one time or another. Before the 1940s, a fair without a horse race was an oddity. Horse racing enthusiasts organized themselves into groups called turf clubs. In many communities the turf club and the agricultural society were so closely related as to be indistinguishable. Often the turf club owned the fairgrounds or vice versa. While horse racing was not exclusive to fairs, it was certainly a large part of almost every fair's program of events until recently.

On the way to the races. This horseman was photographed arriving at the CNE with his two racehorses and equipment. The wagon contains two sulkies and sundry other paraphernalia. The immaculately attired gent leads his two racehorses behind the wagon. It is fitting that leading the whole procession is that invaluable part of any mobile operation of this era: the general-purpose wagon horse.

The English brought their love of horse racing to Canada in the 1700s. The English form of horse racing was thoroughbred racing, in contrast to the American love of pacers or trotters. Everywhere there was an army outpost there were horse races. Halifax, Toronto, Quebec City and Niagara were all prominent garrison towns.

To the refined Englishman, early Canadian horse races were pretty pathetic. A race in 1828 Upper Canada was described by one such visitor as follows:

"Four horses started for a bet of 10,000 ft. of boards. The riders were clumsy looking fellows, bootless and coatless. Before they started, everyone seemed anxious to bet upon some or other of the horses. Wagers were offered in every part of the field, and I was soon assailed by a host of fellows, requesting me to take their offers. The first to attract my notice said he would bet me a barrel of salt pork that Split the Wind would win the day. When I refused to accept this, another offered to bet me 3,000 cedar shingles that Washington would distance every damned scrape of them. A third person tempted me with a wager of 50 lbs. of pork sausages, against a cheese of similar weight, that Prince Edward would be distanced. A fourth, who appeared to be a shoemaker, offered to stake a raw ox hide, against half its weight in tanned leather, that Columbus would be either first or second. Five or six others, who seemed to be partners in a pair of blacksmiths bellows, expressed their willingness to wager them against a barrel of West Indian molasses, or $20 in cash. In the whole course of my life, I never witnessed so ludicrous a scene."[1]

After 1840, Canada imported American thoroughbreds to improve

Early starting gate, Armstrong Fair, B.C., 1920s. This crude method of starting races led to a lot of squabbling and protests among horsemen. It was not only unfair, but dangerous. A Canadian, Wilson McGuinness, invented a device that opened all gates at the same time and hence removed one of the greatest problems with thoroughbred racing.

the quality of its racehorses. Soon, turf clubs and race courses began to proliferate throughout Eastern Canada. Montreal held a King's Plate race in 1836. The king of England, William IV, donated a silver plate worth 100 guineas for the winner. Toronto received Royal Assent to hold a Queen's Plate in 1860, then named after reigning Queen Victoria. In fact, the Queen's Plate is currently the oldest continually run horse race in North America. Finally, the stock of Canadian thoroughbreds improved during the American Civil War (1861 to 1865), when a large number of American thoroughbreds were brought into Canada. Many famous horse breeders, especially in states like Kentucky, shipped their herds to Canada to avoid losing them to the war efforts.

Turf clubs and horse racing gradually spread their way west. Calgary was a horse town from the start. It was only natural that horse racing was added to the fair in 1891. At the Prince Albert Fair, early horse races were run on a challenge basis. The loser lost his horse and saddle. The loss of the saddle may have been the more serious! When that fair got into serious financial difficulties, the turf club bailed it out. In fact, the Lethbridge Fair started out as a big race meet before it became an agricultural fair.

Horse racing itself was a somewhat harmless sport. But the betting and gambling that accompanied the races was another story. Wagering on horse racing soon attracted the attention of anti-gambling lobbyists. Many societies were censured for spending too much time and money on horse racing at the fair. One farmer, so disgusted with the emphasis placed on horse racing at the Aylmer (Ontario) Fair, foreclosed his mortgage on the fairgrounds. And if the fair was hit by financial troubles, race purses were sometimes paid in full before other prize money. In 1894, an Ontario statute banned horse racing at agricultural fairs. The horse racing fanatics had several options. One was to ignore the law. Another was to circumnavigate the law. There were many ways to accomplish this. One was to limit entries to farmers in the local area. These events were called "farmer's trots" or "trials of speed," and placed in the carriage-horse class. One farmer's trot at the Peterborough (Ontario) Fair produced ten entries, only one of which included a certified local farmer. Another evasion tactic was to limit entries to horses not trained on a racetrack. Perhaps this new category created a new type of entry — the free-range race horse. Other fairs substituted road wagons or carriages for sulkies. Another favourite tactic was to show the horse in a class that featured one stint in the show ring and another in a "trial of speed" on the racetrack. When asked what the difference between horse racing and trials of speed were, the superintendent of agricultural societies in Ontario mused: "We call it speeding in the ring when it is our fair and horse racing when it is held under the auspices of some other fellow's fair." [2]

Try as they did, the anti-racing lobby could not eliminate the sport. Racing was a proven crowd-drawer. Halton County Fair in Ontario dropped horse racing from its program for one year, and its attendance dropped by half. The next year, racing was reinstated, and attendance came back to normal. The lesson was learned. A poll of forty-one Ontario fairs in 1900 showed that thirty-six still held some form of horse racing.[3] Eventually, the ban in Ontario was lifted. The Canadian Trotting Association and the Ontario Jockey Club assumed regulation and control of the horse racing industry. The sport became more centralized, with

Standardbred horse race, Kinmount Fair, Ontario, 1950s. Horse racing had passed its zenith at fairs and would soon decline on the fair circuits. Permanent racetracks would gradually replace the annual fair as the hotbed of horse racing. It became too expensive and too dangerous to hold horse races at many of the smaller fairs.

many smaller fairs cancelling horse racing. Over time, horse racing became concentrated at a few larger tracks. And some of these tracks were not even fairgrounds! These new super-racetracks cut into the fair's former monopoly on Canadian horse racing. Many fairs also had to cancel the sport due to growing safety and insurance concerns.

Horse racing in Western Canada never suffered from the same moral dilemma as it did in Ontario. The Prairie was horse country, and saddle racing and chuckwagon racing were a way of life. While betting and gambling remained a concern, it was never a big enough problem to threaten horse racing. In fact, horse racing served as a valuable source of revenue for many fairs. The 1913 Edmonton races netted a gross profit of $6,781, $6,000 of which was given to the city of Edmonton to spend as it saw fit. What politician could oppose such a scheme?

Rodeos and Stampedes

No history of fairs in Canada would be complete without a discussion of rodeos and stampedes. A rodeo is an exhibition of cowboy skills. A stampede is a celebration combining rodeo contests, exhibitions, dancing and more. The word "rodeo" comes from the Spanish word "rodear," which means to go round and round.

As the western North American plains were settled, a cowboy culture was born. It originally started in the United States, but quickly spread to Canada. Buffalo Bill Cody's Wild West Extravaganza was likely the first cowboy show to tour professionally. By 1900, cowboy sports or rodeos were becoming common throughout the Prairies. The Territorial Exhibition, held in Regina in 1895, included a rodeo. The Medicine Hat Fair was sponsoring cowboy sports by 1900, and cowboy contests were held in Calgary for the first time in 1904. At the 1908 Dominion Exhibition in Calgary, Miller Brothers 101 Ranch Wild West Show was a headliner. The first Calgary Stampede was held as a separate event in 1912. It drew 40,000 people, many brought in on special excursion trains from all over the Prairies. However, there were so many organizational problems with this first stampede, that a second edition was not held until 1919.

The founder of the Calgary Stampede was an American cow-

Start of the Wild Horse Race, Lethbridge Fair. Note that some of the horses look wilder than others. In the background is a magnificent Crystal Palace complex: the sign of opulence and prestige at any Canadian fair. The title of the photo reads "Amalgamated Fair and Stampede." Agricultural fairs and stampedes were obvious partners.

boy named Guy Weadick. He was a lover of the Old West. In the early 1900s, Weadick moved to Calgary because he believed this area still embodied the Old West cowboy culture that was now disappearing in the U.S. To promote these waning cowboy ways, Weadick helped organize the first Calgary Stampede in 1912. He was also a prime mover in its rebirth in 1919 and the merger with the Calgary Exhibition in 1923.

The original Stampede needed a lot of fine tuning. The corral was too big and it had no sides. Calves were often roped out of sight of the grandstand, if at all. There were no chutes for the animals to be launched from, so safety was a factor. But there was no problem in attracting crowds or competitors, so the technical issues were quickly solved. Within a few years, stampedes became common all over Western Canada. The Calgary Stampede quickly grew into a premier event in North America. To put it mildly, it was famous. By 1923, Weadick earned the same salary as the mayor of Calgary ($5,000), even though Weadick worked only six weeks a year. Weadick was a master of promotion. In 1923, he was able to rope in the Prince of Wales as a Stampede guest — a real publicity coup. The Prince presented a silver cigarette case to the bronc-riding champ, who refused the gift with the terse comment "I don't smoke." The Prince turned to Weadick, who saved the day by whispering, "Tell him we'll get him a gold watch."[4]

Under Weadick's direction, the city of Calgary turned the Stampede into an event. Local stores put up rough lumber facades and hitching posts. Chuckwagon breakfasts were held on the streets. Cowboys rode their horses right into the stores. Vehicle traffic was banned for two hours each morning on many streets: it was horses only. Calgarians dressed in Old West costumes. Huge Buffalo barbecues regularly served up thousands of burgers. And over 10,000 cowboys and cowgirls danced the night away in the massive annual street dance. Attending the Calgary Stampede became the thing to do for movie stars, politicians and world heads of state. In 1924, Hollywood came knocking and made a movie, *The Calgary Stampede*, on

Saddling a horse for the wild-horse race, Calgary Stampede, 1912. The event looks dangerous... and was! By 1912, stampedes had developed basic safety features such as a fenced corral. No major Western Canadian fair was complete without some sort of rodeo.

Roman race, Winnipeg Stampede, 1913. This race combined speed, skill and danger. How it got its name is open to speculation, although in the days before motion pictures, historical spectacles were a part of many fairs. The lady rider is leading at this point. Rodeos were one place were women were allowed to compete on an equal footing with men.

95

A four-horse chariot race ... of a more recent era than ancient Rome.

location with cowboy actor Hoot Gibson. Fairs just can't buy advertising like that!

By the 1940s the three hottest events in North America were the World Series (baseball), the Indianapolis 500 (motor car racing) and the Calgary Stampede. For the 1950 edition, invitees included the prime minister, the governor general, lieutenant-governor of Alberta, premier of Alberta, mayors of Edmonton and Toronto, three federal cabinet ministers and the leaders of all major federal political parties. All accepted, for it was a must to attend the Calgary Stampede!

Royalty were no strangers to Western stampedes. A special Royal Command Performance of the Calgary Stampede was held in 1952 to honour Queen Elizabeth's visit. And the 1987 Medicine Hat Stampede was visited by Prince Andrew. While meeting a lineup of local dignitaries, he asked one gentleman if he was indeed a real cowboy. Upon the affirmative reply, the Prince said, "Thank goodness! I have met a politician dressed up like a cowboy, a banker dressed up like a cowboy, a druggist dressed up like a cowboy, and a retired dentist dressed up like a cowboy. It's good to meet a real cowboy."[5] Somebody should have informed the Prince that at stampede time, everyone's a cowboy.

With the popularity of stampedes came a few problems. A major issue at the Calgary Stampede for instance, was congestion. When the new midway sponsored a midnight madness special in 1977, over 58,000 people showed up and remained all night to enjoy all the rides for a flat fee. The next year, a study by Stanford University was commissioned to look at ways to relieve overcrowding. The recommendation was to separate the agricultural events from the Stampede. So a separate agricultural show was subsequently held in the spring. The result was a disaster. Both halves of the fair were immediately reunited.⁶ A fair was still a fair.

Stampedes have many events. Some of the more common are bareback riding, bull riding, calf roping, steer wrestling, barrel racing, saddle-bronc riding and wild-cow milking. In later years chuckwagon and chariot races were added. Chariot racing, not really a native Canadian sport, was labelled a cross between chariot battles of ancient Egypt and an open prairie stampede. Chuckwagon races, on the other hand, grew from local Canadian legends. At the end of a roundup of cattle, the ranch cook and the chuckwagon crew would hurriedly load up all their equipment and race other ranch crews to town. The first outfit's wagon to reach the saloon would win, and a round of drinks was usually the prize. Stampede-style chuckwagon racing was a little different. It involved loading a stove in the back of the chuckwagon, racing around a track with outriders on horses, unloading the stove and starting a fire. The first crew to have smoke coming out of their chimney was declared the winner. Chuckwagon racing was a dangerous, exciting affair, often involving spills and thrills.

Chuckwagon races at a fair. Even today, this most popular of rodeo events crosses the speed of horse racing with the thrills of a demolition derby (or chariot race).

The Travelling Rodeo

The travelling rodeo became common all over North America in the early 1900s, especially in the East, beyond the traditional cowboy pale of the prairies. In the summer of 1900, the Oklahoma Wild West and Congress of Rough Riders Show toured Ontario. Their stint at the Peterborough (Ontario) Fair opened with a Wild West parade down Main Street, a well-used advertising ploy to whet the local imagination. A Native American museum on the grounds featured a spirit dance. The grandstand show included the following: Horse Hair George picked up a 50¢ piece from the ground while riding by at full gallop. He then joined La Belle Marie (what stage monikers!) for a trick-roping routine. Two acts followed, with Indians attacking first a settlers cabin and then a stagecoach. Both scenes ended with cowboys driving off the attackers in a noisy display of riding and shooting.⁷ The plot for these touring acts were later copied in countless "Cowboy and Indian" movies.

Bison-powered cart. This team was part of a Wild West show as evidenced by the parade of horsemen. The farmers of Eastern Canada were astounded that the wild buffalo of the prairies could be so domesticated. By the turn of the century, there were only a few thousand buffalo left from the mighty herds that once roamed the Prairie provinces. That fact made this show a display of an endangered species.

Sports Connections

Organized sports have been part of Canadian fairs since the beginning. The first fair in Canada, at Windsor, Nova Scotia, featured wrestling and marksmanship. Athletic events, because they featured local talent, were easy to organize, cheap and readily available anywhere in Canada. Team sports were also a big part of the entertainment package at many early fairs. Tug-of-war, baseball, lacrosse and field sports, now known as track and field, were popular. Rowing races, marathon swimming and powerboat racing were held at fairgrounds that were near water. By the early 1900s, bicycle racing became a feature. Auto racing was also a great crowd pleaser. Organized sports were usually held at facilities already on the fairgrounds. And most fairgrounds already contained a racetrack, but baseball diamonds and lacrosse fields were also often common.

In many smaller towns and villages, the largest buildings or halls in town were located on the fairgrounds. To enhance the fair, most agricultural societies constructed arenas, exhibit halls, barns and display buildings. Ontario even passed a law called the Community Centres Act which gave grants for agricultural societies to construct such public buildings

A very early automobile race. The grandstand is only half-full. I bet it would be full for a horse race!

on their fairgrounds. Naturally, local hockey teams would use these arenas for their games. The arena often did double duty: housing exhibits at fair time, general shows at other times (for example, farmers' markets, conventions, political rallies) and serving as an indoor sports facility.

Of all sports, hockey was clearly the favourite. Some fairs even went into the professional-hockey business. The Ottawa Silver Seven netted Stanley Cups in 1903, 1904 and 1905, playing out of the old Manufacturers Building on the Ottawa fairgrounds. Calgary had hockey by 1919, and the Edmonton fairgrounds became home to hockey in 1913, when the livestock pavilion was converted to a hockey rink. Box seats cost a dollar, cheap seats 50¢ or 75¢ and kids were admitted free.[8] But kids were not allowed into the building until just after the faceoff! The Edmonton Fair Association eventually sponsored the Edmonton Flyers, which won the Allan Cup in 1948 and went on to become a farm team for the Detroit Red Wings. Also operating from the Edmonton fairgrounds were the junior Edmonton Oil Kings, who amassed a total of six Western Hockey League titles and two Memorial Cups.[9] To this day, both Edmonton and Calgary have NHL hockey teams operating from their grounds. The Vancouver Canucks began their amateur and professional (NHL) lives at Hastings Park, the local fairgrounds. And in many smaller centres, the local rink is still usually located on the fairgrounds. To be sure, the outgrowth of professional sports was not restricted to hockey. When the Toronto Blue Jays began their Major League baseball career in 1977, they did so in the old Exhibition Stadium on the CNE grounds.

Many fairs have tried other sports as part of their entertainment package. The Canadian National Exhibition in Toronto has sponsored a variety of these events, including marathon swims and the ever-popular CNE air show, which still packs in the crowds every Labour Day weekend. The Central Provincial Exhibition in Armstrong, B.C., hosted the Interior British Columbia Wrestling Championships. The Calgary Stampede held a ski jumping contest in 1921. The jumpers took off from the roof of the grandstand. Over 10,000 viewers showed up, more likely in anticipation of spectacular spills than for love of the sport. Amazingly, nobody was killed. A Chinook wind cancelled the event the next year, and it was never revived.

Hockey game, Edmonton Fair Arena, 1921. Fairgrounds were a natural place for construction of sports facilities. The buildings could serve multiple purposes. Notice here the skylights in the roof, which allowed maximum use of daylight. Notice also the goal judge standing on the back of the net. This practice was not only dangerous, but a chatty goal judge could be very distracting. Eventually, judges were moved to safety behind the glass, and eventually into glassed cubicles to keep them out of the hands of the crowd.

Start of the marathon swim, CNE. For many years, the CNE held a marathon swim competition as part of its festivities. The event attracted a lot of contestants, many of whom did not finish. Weather conditions and water temperatures regularly wreaked havoc with the race. The first swim, in 1927, was organized following the publicity generated by Toronto resident George Young, who won the Catalina marathon in California. No entrants finished in the women's division, 21 miles, that first year, so the distance was shortened to 10 miles and eventually to 3. In 1928 and 1964, no men finished the race due to bad weather. The last race was held in 1964.

The marathon swim competitions led to some interesting incidents. One young man talked the officials into letting him enter at the last minute. At race time, he smeared himself with black grease, a common precaution against cold water, and lined up at the starting dock. At the sound of the starter's pistol, he leapt into the water and promptly sank like a stone to the bottom of Lake Ontario. A race official, immaculately attired in a three-piece white suit, dove into the water and rescued the hapless competitor from the floor of the lake. The black grease on the white suit made quite a spectacle as the wannabe marathoner was dragged to shore. Of course, the inevitable "What were you doing?" question was posed. The non-swimmer sheepishly confessed he had travelled all the way from Kirkland Lake in Northern Ontario to collect a $25 bet based on whether he would enter the race. At least the CNE officials could be proud that their reputation had spread so far from Toronto![10]

Contests of All Sorts

Recognizing the importance of entertainment to fairs, some directors tried to marry the agricultural sector to the entertainment component, resulting in some novel shows. One exhibition held a corn-eating contest ... for pigs! The participating porkers started out with two hundred kernels of their favourite chow, with additional quantities of one hundred added at intervals. After a five-minute feast, the hog who had devoured the most kernels was declared the winner. Regrettably, this "sport" never became an Olympic event. Another agricultural sport that lasted briefly on the fair circuit was boxing on horseback. It actually resembled medieval jousting, as two riders tried to "unhorse" each other with poles tipped with boxing gloves. It never caught on either. One Ontario fair held an essay contest; the topic was: "Requirements of a good farmer's wife." The essays were read aloud at the fair by two young farm women. It was not a good spectator sport. Some pre-1900 agricultural societies held turnip-hoeing competitions. One match attracted sixty entrants! Other agricultural entertainments have enjoyed better popularity. Cow milking and sheep-shearing contests are still found at many fairs. "Farmer's Olympics" are also popular. Tractor pulls allow local farmers to show off both their skills and machinery. Sports involving saddle horses and pets are likewise very popular today.

Another popular contest was the baby show. Anyone who has ever judged one of these can relate to its perils. There is never a clear consensus on how a "first-place baby" should be judged: looks? health? size? interaction with others? Of course, every parent and/or grandparent maintains the opinion that his or her entry is number one. Disapproval with the judges' decision is expressed in a variety of ways, and it has been wisely suggested that judges for the baby show be from out of town — and get out of town quickly after the show.

Calgary Fair sponsored its first baby show in 1889. At the start of the contest, the judges noticed two things. First, the room was filled with excited mothers and restless fathers. The mothers sat with the little contestants while the fathers stood impatiently around the perimeter of the room. The second rather alarming fact was that most of the fathers were armed. In cowboy style, the young dads had worn their six-shooter pistols to the fair. Panic gripped the judges as they imagined the worst. Disappointment with the results could bring a hail of lead from the audience. What to do? One quick-thinking director realized the sheep show had been a bust, and a number of first-place ribbons were left over. The judges went through the motions, declared the contest a tie, gave all the contestants a first-place ribbon and breathed a collective sigh of relief. The baby show was not held the next year.[11]

These anxious mothers and their happy babies are not here for the firearms demonstration but the baby show, CNE, 1960s. The children have no idea of the pitfalls of stage performances.

Interestingly, the baby show was once a form of public education. Health officals sponsored the show and used it to promote public health and hygiene. The judging of one such show was described as follows:

"Every available chair in the tent was occupied by a mother intent on removing the clothing from a squirming little bundle of humanity on her lap, who insists on throwing his arms and legs in the air and will put his hand in his mouth just when she wants to slip off the sleeve of his dress. There are dark babies and fair babies, babies with fluffy little curls and babies with wee bald heads. Some are clean and gurgling with delight, others are raising their tiny voices in protest."

A nurse awarded points for clothing, and deducted marks for uncleanliness or defects. The baby was measured and weighed. Additional points were given for nutrition, skin condition and habits. This last category was no doubt a controversial one. The above observer noted: "Several cunning wee tots lost a point or two because they would suck their thumbs."[12]

The Grandstand Extravaganza

One of the key features of fairs has been the stage or grandstand show. No fair would be complete without some sort of stage show. The earliest midways put on small stage shows, but these were different from the grandstand-style show. The grandstand show was presented in front of a larger crowd, usually seated on the fair's grandstand, of course. Sometimes a separate admission was charged, but for most fairs, the grandstand show was free with the price of admission to the grounds. Grandstand shows have varied throughout the last hundred years. The earliest, which emerged around 1900, were variety shows, usually involving animal tricks, circus acts, and clowns or vaudeville-style performers. These acts were usually contracted directly to the fair board by an agent who booked on a ciruit basis. In the days before movies and television, the grandstand show was the height of the entertainment year in many smaller centres across Canada.

"Wild Men from Borneo," a sideshow circa 1910. The foreign and exotic were a big draw. Most Canadians had no idea where Borneo was, let alone what the "wild men" looked like, so it was easy to dupe the public. However, these dancers certainly look authentic.

These early acts often incorporated performers who went by embellished names. The moniker "Professor" was often used. Professional and military titles such as "Captain," "Major" or "Colonel" added a touch of class to performers. Oddly, "General" was only rarely used, perhaps a sign that organizers were unwilling to push the limits of credibility.

Odd and humorous names appear again and again in early advertisments. The 1901 Edmonton Fair, for instance, contracted with the Minneapolis Amusement Co. for the following show:

The Richards Family: aerial artists
De Vallas: horizontal balancing act
Professor Bell's Japanese slack wire act
Professor Bell's jugglers
Professor Norquist: high diver
Matthews and Newman: unicycle act

Foreign exoticism was also a good sell. In 1908, Campbell Brothers Consolidated Show presented the following lineup:

South American gauchos (cowboys)
Australian boomerang throwers
Bright eyed senoritas from Mexico
Desert-born Bedouin Arabs
Japanese athletes
Hindu fakirs
Cavalry troopers of many flags
Strange oriental music
Funny clowns with funny mules
Mexican, Indian and American bands
Master horsemen of the universe
Wild beasts from the desert [caged of course]

Wow! What a lineup for the bored farmhand or cowboy!

Professor Gentry's Travelling Animal Circus Show, CNE, 1900. Trained animals were always popular entertainment, especially in the era before the spread of mechanical rides. In the centre is a real horse acting like a rocking horse. An odd but alluring group of unidentified creatures are fenced in the centre. The high-wire pole leaves a great deal to the imagination about what act will happen there.

Another outfit playing the Calgary fair featured:

Golems: 8 acrobats from the court of the Shah of Persia
Howards' dogs and ponies
C. W. Parker Carnival
A. G. Barnes Wild Animals Circus
Three-girl motorcycle act from New York

Smaller fairs naturally had smaller grandstand shows. Some of these included:

Marvelous Marsh the diver
Kelly and Ashley: elastic billiard table
Three Zenos: gymnasts
Austin Sisters: trapeze artists
Fowler Brothers: bicycle troop
Nelson Sisters: aerial ladder act
Flying Jordan family

**Clara Beckwith's Underwater Act [eating, drinking and sewing underwater]
The Ringers: high wire act
Toyama: Japanese balancing troop
Cervene dancers
Jesse Blair and the Glasgow Maids
The farmer and his pigs
South American grave robber: an armadillo act
The human cannonball**

Poster promoting the Wulfronas Acrobatic Troupe from Paris at Brandon Fair. Such acts were common at fairs. Exotic homelands added to the mystique of any act. Many acts used terms like "by appointment to his Royal Majesty of…" (insert foreign country here). Unfortunately, in recent years the world has began to run out of "Royal Majesties."

Titles and enchanting names certainly piqued the imagination! Neromus, the bull wrestler, for instance, appeared in 1904 at the Winnipeg Fair. The crowds were so large that two hundred fair-goers scrambled onto the roof of a cattle barn to gain a better view. In the interest of safety, a constable was sent to remove them. Unwilling to leave, the watchers crowded to one end of the shed, which promptly collapsed under their weight. Thirty-seven were injured. One man, trapped between two excited bulls, was asked, "What can we do to help you?." His reply was, "Tell me who is winning!"[13] Apparently, the show must go on!

These variety shows were booked either at the provincial fair conventions, or more commonly, by simply showing up at the booking agent's office in the nearest large city. In many situations, the lineup was not booked very far in advance. A committee would often travel to see the agent and find out what was available mere weeks before the fair started. The booking agent would piece together a show from what was currently available. The whole operation was rather casual, and variety was the name of the game. The entertainment committee from the Peterborough (Ontario) Fair, for instance, was given the following list of performers to choose from by their booking agent:

**Dutchy, the trick mule
La Tow Sisters: drama
La France trio: comedy
Miss Dorothy DeVonda: lady aeronaut
4 Mayos: skating comedy routine**

Miss Delora: the physical culture girl
Japanese wonders: jugglers
Choy Gar Duo: knives
Sig Franz Duo: trick cyclists
Demarlo and Marlette: trapeze contortionists
Emilio's Royal Doberman Pincers
Hanneford Family: circus equestrians
Five Juggling Jewells
Pallenberg's Bears[14]

All this variety made for some wild and impressive grandstands. Fred McGuinness, entertainment director for Brandon Fair, described one such show in the 1950s as follows:

"The grandstand show was a creature of the night, the darker the night the better. Only when the lighting could be controlled was the night show at its best. I once viewed it as a waste of time when we felt we had to run grandstand matinees for the kids. Members of that audience faced north and within their view was a vast panorama: a tiny stage on the verge of a racetrack, an infield filled with trucks and trailers, a long row of a race-horse stables, a rail line, the skyline of the city. A juggler or a balancing act had to be good indeed to capture attention in such a setting.

"Even in the evening, when that long, slow, colourful Prairie sunset is at its most spectacular, the first half of the show often failed to capture the public's attention. The lighting was simply ineffective, the stage settings subdued; even the sequins had no sparkle.

"Only in total darkness could the night show live up to its name and become a grandstand show. Only in the out of doors blackness could show business display its magic, the ability to transport you and your imagination into a czar's court, or a fairy kingdom. In those happy days of the early 50s, we had such a demand for tickets that we staged two complete performances every Thursday and Friday.

"In the 30s, and 40s, and for almost exactly half of the 50s, Prairie audiences appreciated the grandstand show for what it was, a means of transporting us into a different galaxy. For two hours every year patrons forgot about manure forks, mortgage payments, and those other exigencies of farmstead and homestead.

Elephant rides, Edmonton Fair, 1919. An exciting adventure for Canadian children of any era is an elephant ride. The huge, lumbering pachyderm could carry eight children at a time. Fairs were the place to see the exotic, the unusual and the unknown from all over the world — at least until TV came along.

A public hanging, Calgary Fair. At least we hope it was a mock affair. And no, this was not the fair president after a bad stage act. We are not sure how or why this act would be popular, or what fair-goers hoped to see. Such hangings were not an everyday reality in the Canadian West. It does prove the old adage however, that fairs will try anything to draw a crowd!

"Each year we were quite hopelessly spellbound as act followed act. Right on cue we gasped, chuckled, sucked in our collective breath, and roared out those belly laughs. It was even possible for some of us over the years to trace connections from one aspect of show business and another. One evening as I sat with 5,000 others and watched a slender teenager fired from a cannon by her father, Alfred Zacchini, I would never have guessed that in ten years she would be back on the 'A' circuit in another capacity, the brand new bride of Carl Sedylmayr Jr. (Royal American Shows).

"The M.C. then gave us the theme, which usually had one foot in ethnic history and the other in some global undertaking. This done, however, it still was not time for the opening act; the master of ceremonies had to give his inventory of adjectives their annual work-out, which he did with a catalogue of acts and performers. This year's parade of talent was 'unsurpassed.' The aerial acts were 'death defying.' The animal acts were quite clearly 'world renowned.' The soprano and the baritone had sung command performances before 'the crowned heads of Europe.'

"Prairie audiences were called 'red bloomer' crowds. A 'red bloomer' crowd is a farm audience. Once each night they want to see the clown do something which causes his suspenders to break and his baggy pants to fall down. I remember the last year this happened while he was hanging from a high wire. Well, when this happens the audience learns that he's wearing women's knickers, that were grossly oversized, and always, always bright red. Last year, those bloomers gave the crowd the biggest laugh of the evening ... And it worked every evening from Brandon right through to Port Arthur and Fort William six weeks later. The crowd likes things done the same way every year.

"Let us not forget that clown or those red bloomers, for the clown was a key part of the show. He provided the comic relief which was mixed in with music, human talent, animal talent, balances, surprises, apprehension, smiles, and thrills which allegedly defied death.

"The headline act was as functional as the heading in your newspaper; its duty was to attract attention. It was the second-highest paid act, second only to the one which closed the show. The opener set the pace, set the theme, elevated public expectations that, for the next two hours, there would be exposed before them their entire annual intake of professional entertainment. Farmers, folks from the small towns, city residents, wives and mothers and over-tired kids gathered nightly in batches of 5,000 for the annual 'extravaganza.'

"I saw my first TV in 1952. Never could I have reckoned at that time that this great leap in electronic communication would sound the death knell of all our connection with outdoor show business. The availability of entertainment in the living room all day every day spelled the end of these delightful open air, summertime presentations."[15]

Winter Scene from "Banff."

The Pageant

Another branch of the grandstand show was the pageant, an elaborate stage show, which was produced locally, and which usually celebrated a historical theme. Pageants involved huge, complicated sets and large casts of mostly amateur actors. Casts of more than one hundred actors were common. Due to the size of the pageants, they could not be staged in conventional theatres or auditoriums. The pageant was the child of the

"Down on the Farm."

107

"Gunfight at the OK Saloon."

open-air stage or grandstand show. The fair and the pageant were natural allies. Most of these outdoor stage spectacles were produced at the larger fairs. They needed big acts to fill dates for their grandstand-show lineup. Often special effects such as fireworks, cannons, live animals, entire bands, fires and special lighting were employed. It was electric lighting that enabled the pageant to become an art form.

The CNE staged its first pageant spectacle in 1884 with the "Bombardment of Alexandria." Warm-up acts consisting of acrobats, trained animal shows and music preceded the main attraction. Elaborate fireworks and lighting displays replicated the battle and burning of the famous city in Egypt. In the era before motion pictures, such a spectacle was truly special for fair-goers. The CNE staged a pageant every year between 1884 and 1941. Some of the shows were:

The Siege of Peking   **The Siege of Algiers**
The Siege of Sebastipol   **The Siege of Lucknow**
Burning of Moscow   **Empire Triumphant**
Last Days of Pompeii   **Burning of Rome**
The Siege of Paris

Canadian themes were emphasized after World War I with such titles as:

Under the Maple Leaf
In Far Away Calgary
Sergeant Ferguson Northwest Mounted Police
Birth of Canada[16]

The larger fairs in Western Canada also produced pageants. In 1926, Edmonton staged a huge historical pageant to mark the twenty-first birthday of Alberta. The 1927 pageant had a cast of 1,000 performers, which was 1.3 percent of Edmonton's total population. Pageants featuring Native peoples were often staged at the Calgary Stampede. The Battle of Batoche and the North-West Rebellion were historical events frequently re-enacted. Saskatoon Fair produced the "Pioneers of the Prairie" and the War of 1812 during the 1920s. But by the 1930s, the days of the pageant were numbered. They were labour intensive and expensive to produce. Only very large fairs could afford such spectacles. Radio, and later television, led to the rise of the musical superstar. Stage shows changed from casts of hundreds to a few big-name performers.

Bandshells and Big-Name Stars

Musical performances have always been a big part of fairs across Canada. In the days before sound amplification, musicians relied on sheer numbers to get their sound out across the crowds. Large bands were the order of the day. Many fairs had a band concert as part of their grandstand lineup. In the early 1900s, marching brass bands were popular: every town of any size had one. In 1878, fifty-one Ontario fairs had brass bands play at their exhibitions. They marched in parades and played concerts as well. It was only natural for this local entertainment to perform at the neighbourhood fair. Where available, military bands were also included. Stages where they played were called bandshells because large bands were the primary acts. In 1919, John Philip Sousa, the dean of American brass-band leaders, played the Edmonton Fair and helped set an attendance record.

In the years following World War II, big-name stars began to emerge in the music world. Radio certainly advanced the music industry, and later television did its share. Pageants disappeared and brass bands were sent back to the parade. Music stars such as Danny Kaye, Andy Griffin, Anita Bryant, Andy Williams, Red Skelton, George Gobel, Wally Costner, Tommy Hunter and the "Country Hoe-Down" featuring Gordie Tapp dominated the grandstand shows. These stars all had roots in the new mass media. You could watch Jimmy Durante on TV and listen to Doris Day on radio, then see them live at the larger fairs. Different types of musical acts started to appear. The CNE stage shows would include something for everyone, with performers ranging from Perry Como to Pat Boone to the Beach Boys and The Guess Who. Just about every country 'n' western performer you can name performed at a fair.

Meet the band! The Coldstream Guards in a very martial pose.

Morning concert by the Coldstream Guards, a famous British army band, CNE, early 1900s. Obviously a popular attraction, this famous British military band toured extensively in Canada. In the days before musical superstars and electrical amplification, bands such as these were the big attraction at many fairs.

Larger fairs began to rely heavily on big-name stage shows to attract large crowds. The 1971 Pacific National Exhibition (PNE) lineup contained the following name acts: Tom Jones, Ray Charles, New Seekers, Jimmie Rodgers, and for Canadian content, Anne Murray, the Irish Rovers and The Royal Canadian Mounted Police Musical Ride. The size and affluence of a fair was often judged by its stage show. The desire to have a big-name grandstand act dominated the agenda of fairs big and small. "Who's playing the fair this year?" or "Who's on the grandstand?" became common inquiries.

The role of entertainment at local fairs was a contentious issue. The dependence on the stage show worried many fair directors because they felt fairs were supposed to be much more than grandstand shows. Yet, many fairs, especially the larger ones, felt these grandstand shows were needed to get patrons to the fair. The subtle change in advertising mirrors this change in fair priorities. A hundred years ago, fair posters proudly proclaimed the sum offered in prize money, while today's posters emphasize the entertainment lineup. The agricultural component is still alive and very vibrant. It's a fact that most people want to be entertained as well as educated. Or as one fair-goer I once spoke to said, "I know I can see livestock at the fair. But I want to know who's on the stage show."

Strobel's airship, Calgary Fair, 1908. One of the highlights of the Dominion Exhibition of 1908 was this airship or dirigible, which was filled with hydrogen. The passenger gondola looks primitive and downright dangerous. However, the airship made five flights during the fair, all successful.

Entertainment From Above

Hot-air balloons were a popular novelty act at fairs in the early 1900s. In the era before the airplane, the hot-air balloon and its cousin, the airship, had a monopoly on airborne travel. But the hot-air balloon show was not without its dangers, which, of course, added to its appeal. They were susceptible to bad weather, particularly wind. While launches were difficult, landings were even more dangerous. At the 1888 Ottawa Fair, a member of the ground crew did not let go of the tether rope and was carried aloft by the rising balloon. After a few minutes of dangling from the ascending balloon, he lost his grip and plunged to his death in full view of the spectators. Such hazards reinforced the reputation of hot-air balloons as a dangerous, and therefore must-see attraction.

No, this is not a new ride. You cannot ride the merry-go-round on the ground level and the airplane on the next. Captain Fred R. McCall was giving an airplane demonstration when he developed engine trouble. He chose to crash-land on the midway instead of the racetrack, where an automobile race was in progress. Early airplanes flew so slowly, landings such as this were possible without great damage or death. In today's world, such an incident would have resulted in a massive crash site and great loss of life. Nobody was hurt in this 1919 Calgary Fair incident. It was not recorded if any merry-go-round patrons asked for a refund.

Recollections of The Airplane Ride, 1916

The date was September 14, 1916 — the day was misty and foggy—not a good day for flying! The pilot and owner of the little old plane was one of two brothers of the Colley Air Service out of Trenton, Ontario who had purchased a couple of discarded craft from the Canadian Air Force to conduct a series of what they called "barn storming" performances. They landed and took off just east of where the cattle and horses were tied on the fairgrounds. It was not a perfect landing field with cross furrows every fifty or sixty feet, so that every time they landed they had to tighten up all the turnbuckles, strutts etc, before taking off again.

However, I guess I was of the adventurous type so I spent the last $10 I had for a 10-minute ride. The rider before me was Wm. Bonhower and after me was Arch Tobin, who by the way, reached over and felt the seat of my pants to see if they were dry! They were!

In the meantime, my mother over in the fairgrounds was looking up at that wonderful flying machine circling over the grounds. Someone said to her — "It's your son up there!"

Well I enjoyed it (the ride) and after a lot of severe reprimands, survived it.

P.S. The next day at Chesterville, the old plane crashed killing both the pilot and his passenger.[17]

Air show, CNE. The airplane fly-past began in the 1950s and became a highlight of the last weekend of the Exhibition, held every Labour Day. Planes and pilots from around the world gave this show an international flavor. The extravaganza was not without tragedy: as on at least two occasions, airplanes have crashed, with resultant loss of life.

Many ballooners wore parachutes, not exactly a ringing endorsement of their ability to land safely. At the 1906 Saskatoon Fair, a balloon act was brought in to liven things up. It took three days for the balloon to be inflated with smoke. When ready, it promptly flew a couple hundred yards before crash-landing in the CN rail yards. The smoke escaped and the show ended abruptly, and oh yes, the balloonist parachuted to safety before the crash.

Another feature attraction at early-twentieth-century fairs was the "airplane" demonstration. In the first three decades of the century, the aeroplane was a novelty. Most Canadians got their first glimpse of the winged wonders at a fair. Small or large, fairs throughout Canada brought in airplanes as special attractions. The planes sometimes gave rides, but mostly they just flew overhead or were on display for the curious. The early planes were often brought in by train and assembled at the fair, which was likely a show in itself! The Kinmount (Ontario) Fair contracted with the Colley Aeroplane Co. of Toronto to provide plane rides for their 1919 fair. The cost of the one-day show was the princely sum of $400, but an attendance record was set as curious fair-goers flocked to the fair to see their first flying machine. The CNE, always an innovator, sponsored the great Toronto–New York Air Race in 1919. Contestants were to leave either site and return, with the flyer with the best time over the 1,150-mile course declared the winner. The pilots, and the planes, were a curious assortment, mostly World War I leftovers. They included the famous Billy Barker, V.C. A number of pilots crashed or made forced landings throughout the race, but amazingly, no one was killed. A Major Simmons had a most hair-raising experience while landing in Toronto. A smoke signal, sent up to guide him in, blinded him instead. On his approach to the landing strip, the wing tip of the aircraft dipped so low that it caught the harness of a horse pulling a wagon and tore the harness right from its back. Miraculously, the horse, its driver and both pilot and airplane escaped unscathed. The race was won by a Major Shorty Schroeder, an American.[18]

A less common aerial show was the barnstormer or stunt show. Whereas airplane demonstrations were relatively mild and somewhat harmless, the stunt show had a real element of danger front and centre. These daring young men and women performed various stunts such as wing walking, parachuting and other gravity-defying feats. The barnstormers came from both sides of the border and most had been fighter pilots in World War I.

Canadian barnstormer Dick Granere had a rather unique signature act. Two 40-foot-wide wooden frames covered with fabric were set up. He flew his plane through the first target, did a loop and a roll and then sliced through the second target. Timing, and weak fabric, were the key elements. American captain Frank Frakes added another rather unique twist to stunt flying. He built a house-like structure and then deliberately crashed his plane into it. The house was built to disintegrate on impact, but the element of danger was still very real. Believe it or not, Frakes retired peacefully after his one-hundredth crash. But not all stunt pilots were so lucky. An American newspaper blandly reported in 1926 that a dozen aviators had been killed performing various stunts over the past year. While the stunts themselves were big crowd-pleasers, no doubt many attended hoping to see some sort of crash. One barnstormer retired at the tender age of 25 when he finally realized it was the possibility of his injury or death that titillated the crowd most. "Some of them even resent it if you're not actually killed," the daredevil recalled. However, the same pilot evidently missed the sport, for he was killed performing a new stunt only three years later. The toll of destruction and death in stunt shows eventually led to government regulation and the virtual demise of the aerial stunt show at fairs.[19]

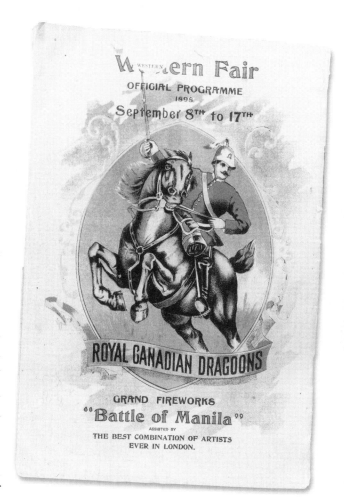

Another fine form of entertainment from above were fireworks, a really big crowd-pleaser at early fairs. They were widely used in such pageants as the Bombardment of Alexandria, the Burning of Moscow and the Last Days of Pompeii. Professor W. T. Hand, of the famous Hand fireworks, made his reputation by staging pyrotechnical spectacles at many fairs. The first Dominion Exhibition at Ottawa featured Professor Hand and his pyrotechnical wizardry. Saskatoon Fair had a fireworks display that featured the following:

Portrait of King George V　　**Union Jack flag**
A lightning thunderstorm　　**Niagara Falls**

While attractive, fireworks displays were expensive. Small fairs found the price out of reach and many larger fairs also had to cancel the displays due to high costs. Still, it's not uncommon to see one of the larger fairs close out for the season with a bedazzling display of sky-bound pyrotechnics.

A water-skiing demonstration at the CNE. In a bid to broaden its appeal and utilize its frontage on Lake Ontario, the CNE for many years held a parade on Lake Ontario called "Aquarama." It featured everything from floats, boats, swimmers and water skiers.

The Changing Face of Grandstand Shows

Major stage shows have several drawbacks. One major problem is their high cost: they are expensive to produce and the costs increase with the status of the entertainers. This cost factor has placed big-name performers out of reach of both smaller fairs and many larger fairs as well. Competition from other venues has also become a factor in some communities. Once upon a time, the local fair was the only venue in town for stage shows. Now theatres, concert halls, arenas and other outdoor venues feature star performances. Live concerts have become common in both cities and towns. Many fairs have been unable to compete.

Some of the larger exhibitions did hold non-fair concerts. Their success was variable. The Pacific National Exhibition had two fine facilities on its grounds: Empire Stadium and the Pacific Coliseum; both venues have hosted a wide variety of events. In 1957, a young singer named Elvis Presley attracted 22,000 fans for a concert. In the words of one director: "Elvis rocked the kids into such a frenzy of joy, that they nearly kicked each other to death."[20] In the aftermath, it was decided that Elvis and his "wiggling songs" be banned from the fairgrounds. Clearly the ban only applied to Elvis, as the PNE later hosted the Beatles (1964) and the Rolling Stones (1972).

To replace big-name stage show, fairs became creative. Talent contests or amateur shows became widespread. Wannabe local stars were invited to strut their stuff for a fraction of the cost. Tribute acts or impersonators have proliferated in the last few years. These imitators try to look and sound like their more famous namesakes, again at a fraction of the price. Less expensive stars are booked by many fairs. And increasingly, fairs have turned to non-musical acts to fill out their grandstand schedules. Demolition derbies, tractor and truck pulls and automobile thrill shows are just a few of the newer forms of entertainment taking place in front of the grandstand. Fairs survive because they are adaptable.

Demolition derby, Brandon Fair. Demolition derbies are crowd-drawers at today's fairs.

SMALL FAIRS VERSUS LARGER FAIRS: IS BIGGER BETTER?

"The Canadian fair system is too successful." This kind of statement has been made many times by people called "reformers" who have tried to change the ways fairs now work. The original purpose of agricultural fairs was to educate and inform the rural populace. However, over time many fairs began to include lots of different types of entertainment. To the "reformers" or "agricultural purists," this meant fairs were straying from their original purposes. As fairs became more deeply embedded in Canadian culture, agricultural purists and government officials were continually trying to bring fairs back to their educational origins. These people recognized the value of fairs and agricultural societies to Canada. They believed this arm of agricultural improvement and education could be made even better with a little help. And of course government officials were always trying to get more value from public funds. Agricultural purists wished fairs would concentrate solely on agricultural improvement and education.

One method of better achieving these goals was to hold fewer but larger fairs. The reformers believed this would provide many advantages. At larger fairs, bigger prizes could be offered, hopefully attracting better-quality entries. It was also believed that too many small township or community fairs cut into the grants and profits of others. Smaller fairs cut into grant money, were poorly organized and did little to accomplish and meet their goals of agricultural improvement and rural education. Thus arose a country-wide debate over whether Canada had too many small fairs.

This debate spawned a lot of soul-searching among Canadian agricultural societies. Different provinces had different ways of addressing fair reform. In 1990, the Ontario Association of Fairs surveyed its members on the issue. In that year, 371 fairs were held in the province: 96 county or large fairs and 275 township or small fairs. The fewer-fairs advocates argued thusly:

"Prizes at township fairs ... are so small that people think it's not worth their while preparing articles for competition, and time is wasted attending the shows."

"At the county [larger] fair ... larger prizes could be given thus inducing greater competition ... the object of which fall fairs were intended, mainly the breeding and production of first class stock and produce; they would be attended with much larger success than at present."

"Those small fairs are kept alive by interested parties who make a point of raising a few roots [i.e. turnips] and fitting up a few animals for the simple purpose of gobbling up the prize money."

A fair director from Teeswater, Ontario, moaned "that 20 fairs were held within a 16 mile radius of his village. He maintained the area would be better served by three or four fairs."

But the small fair was not without its supporters as well:

"If the only object of the fair were to give the farmer and his friends a holiday [in too many cases almost the only holiday in the year] and a chance to compare notes and discuss the year's results, the fairs would be worth all they cost."

"Removing the country fair is equivalent to removing the public-school in rural districts," warned another rural resident. Another called the local fair, "The only school we could afford to attend." One writer wisely pointed out the small, local fair served as a starting point for new exhibitors. Almost all animals and exhibitors at the larger fairs had been previous winners at a small fair.

These supporters of small fairs lined up to take shots at the larger ones, which were regularly criticized for concentrating on non-agricultural events such as midways, horse racing, sports and circuses.

A critic of a city fair wrote.

"The fair seems to be trying to unite an agricultural show with a circus, and the greater attention appeared to be paid to the entertainment part. These larger shows are held in villages and towns and the principal object seems to be to draw a crowd."

A director from Grey County, Ontario, complained about his county fair: "Our agricultural associations are too often controlled by the sporting class. This is especially true where there is the largest town in the electoral district. In North Grey most of the directors are residents of Owen Sound, chiefly sportsmen, and consequently there are three days ... of horse racing, intermixed with dancing and other sports. This is simply disgusting."

Another knock against fewer, larger fairs was simply distance. In the era before motorcars, it was often difficult to travel long distances and back in a single day. Cutting out the local fair meant most farmers and fairgoers had to travel farther to attend a fair. Distance and difficulty of transport would keep many attendees and exhibitors from going to any fair. And finally, community pride was a big reason small villages and townships started a fair in the first place. Having an agricultural show in your community was a symbol of prestige and affluence. To lose that fair was a step backwards for the community. It was for this reason that small villages and townships guarded their fairs with fierce devotion and jealousy. Any attempt by government or its agents to remove fairs from small communities was a politically unsound manoeuvre, especially come re-election time.

Having lost their battle to cut the number of fairs, reformers turned to the more politically sound practice of improving the existing ones. Or as one fair director succinctly put it: "We don't need fewer fairs, we need better fairs." In this field, the reformers enjoyed greater success. Ontario designated seven fairs as model fairs. Directors from fairs across the province were encouraged to visit the models nearest them and learn. Renfrew, Richmond, Whitby and Norfolk were such designates. Essays describing how a model fair was organized were distributed across the province, and prize lists were sent to every fair in Ontario.[1]

Fair reformers in Alberta tried a different approach. Alberta attempted to grade fairs against what it thought an ideal agricultural fair should be. Departmental judges were asked to evaluate the shows using scorecards drawn up by the Ministry of Agriculture. Alberta fairs were then rated on the following:

1. livestock shows
   number of exhibitors } 70%
   good management
2. community involvement 20 %
3. women's exhibits 10%

Most judges were livestock judges: not the best ajudicators of shows involving animals. The Three Hills Fair, for example, complained about its low score in 1922, maintaining that the domestic science and ladies' work were inspected by a judge who did not know the difference between crochet and embroidery. In another instance, the Wainwright Fair was tagged in 1922 as unfit for government grants due to its previous year's poor showing. Oddly, the Alberta judge didn't even factor in the fact that the Wainwright Fair had been rained out that year. The Byemore Fair protested its exhibit hall score, charging that the livestock judges never even viewed their efforts, but instead simply concocted a score. No wonder that disagreements between judges and directors over the quality of Alberta fairs were common.[2]

Saskatchewan tried another approach to reforming its fairs. The government of Saskatchewan attempted to embarrass problem fairs into improving themselves. To this end, Saskatchewan agricultural societies were asked to classify themselves under one of the following headings:

1. Dead societies that appear to see their whole duty to the community in supporting a holiday of horse races and sports which they call an exhibition.

2. Societies that give a little support to the livestock industry towards purely material ends.
3. Societies that give general support to such activities as plowing matches, standing-crop competitions and seed fairs, without any very clear idea of their real value.
4. Societies whose officials see the possibility of substituting craftsmanship for drudgery in all agricultural operations, and recognize in the society's activities the means of bringing this home to the people.
5. Societies whose officials believe that their main business is to foster conditions tending towards a more perfect rural civilization.

I wonder how many societies picked number one?

Time and circumstances solved the problem of the number of fairs across Canada. The advent of the motor car shrank distances and travelling time. Electricity enabled fairs to operate after dark, and the decline in the number of farmers and farms during the Great Depression gutted many rural communities. All these events helped put an end to many fairs. Some societies amalgamated and others just disappeared. But strong, well-managed, aggressive fairs adapted to change — and survived.

TOM BISHOP
4B Ranch Wild West & Rodeo Shows: A Recollection

Our connection with Ontario fairs started around 1912 when my dad, Thos. W. Bishop, began presenting his trained-horse exhibitions. Then, during the First World War he appeared at many Southern Ontario fairs to present his spectacular trained horse Saladin (or Sandy), who climaxed his act when, by voice control only, he mounted a narrow 6-foot-high stage by way of steps, crossed that, mounted a pedestal, and pulling a bobbin, shot a double-barrelled shotgun at an effigy of Kaiser Wilhelm of Germany (the world's villain at that time). My dad went on to produce large-scale Wild West productions of the type of show Buffalo Bill did. (He had seen Buffalo Bill twice as a youngster, to much acclaim and packed grandstands). However, nothing brought the house down like Saladin shooting the Kaiser.

Today we carry on the Wild West tradition, presenting both acts and entire Wild West shows in Canada, the U.S. and abroad. My son, Tom III, carries on the business along with my two daughters, Sarah and Sally, all of whom are trick riders & performers and recently came back from a tour of the Middle East.

Playing fairs, I've found, demands a maximum amount of flexibility from the showmen, and unexpected problems seem to be the norm, along with their inconveniences.

As you would imagine in over ninety years and three generations of playing fairs there is a basketful of incidents and recollections, many of them amusing.

For several years we presented our rodeo at the CNE Coliseum, with bull riding being the best event. One time, a fair employee prematurely opened the huge doors behind the bucking chutes, allowing a gush of fresh air into the jam-packed Coliseum. One of my Brahma bulls thought that "out there" was a better idea than "in here," and jumped over two five-and-a-half-foot fences in the bullpen, out the Coliseum door, across the parking lot, under the Gardiner Expressway, and found the only opening in the thousand feet of security fence. This hole in the fence led onto the CN–CPR railway right-of-way, with at least ten rail lines, in a darkness that was absolute. Some of these lines were for shunting and others for express trains. This caused a shutdown of the rail system for ninety minutes in Toronto, while a dozen of my cowboys, some on foot and some on horseback, frantically searched for my Brahma bull.

Being riled up with all this chasing, the bull suddenly turned and decided to go on the offensive. The resulting scene was hilarious, with seven rodeo cowboys taking refuge on top of one tiny signal box, where they waited to be rescued one by one by cowboys on horseback. We eventually roped the bull beneath a hugh Canada Packers billboard, which along with newspaper photographs, made for great publicity.

With the stories and pictures (and even Gordon Sinclair on CFRB), we got great publicity (a press agent's dream), but alas for the showman's luck, that had been our closing performance.

The following day, I found out directly that city police, railway police and CNR officials were not nearly as amused at our misadventures as the many newspaper photographers, reporters and members of the press!

Yes, there are always challenging times, but humorous and good times too, as we continue to look forward to the wonderful relationships and warm friendships we've had with agricultural fairs in four provinces and six states over ninety some years of fairs in our family.

Commercial Exhibits

AS YOU'LL RECALL FROM THE EARLIER SECTION, medieval fairs were commercial fairs or trade fairs for the buying and selling of goods. While their North American cousins were supposed to be agricultural-improvement fairs, the commercial side was not avoided. Local businesses could not pass up the annual opportunity to display their wares in front of the largest assemblage of potential buyers the community could muster. Until recently, no other single attraction could produce the crowds that the local fair did. So, local retailers, manufacturers and businesspeople used the fair as a marketing tool. The larger fairs and industrial exhibitions attracted huge manufacturers' displays. The latest in fashion-trends, tools, gimmicks and labour-saving devices were all attractively placed before consumers. Fair-goers flocked to the commercial displays, to see the latest trends and inventions. Fairs became huge department stores; no, make that super-malls, where everything was on display. To the housewife, the fair introduced wondrous new appliances and gadgets, such as washing machines and vacuum cleaners. The CNE called one of its commercial exhibit halls the Better Living Building. To the child, amazing new inventions like radios and televisions were first encountered at the fair. And the farmer could inspect the newest in machinery, cars, gadgets and foodstuffs, all on display at the fair.

Canadian Pacific Railway display at a Chicago trade fair. This display is designed to promote the Prairies as a settlement destination. The CPR not only carried prospective settlers to the West, but also sold them farmland when they arrived. Also featured were animals and fish displays, clearly aimed at the tourist trade.

Canadian manufacturers also went abroad to promote their goods. World's fairs, state fairs, international and industrial trade fairs were all excellent venues to promote the newest and best from Canada. International events were also great places to promote Canada as a place to live. Advertising for immigrants was not a new policy for Canadian governments. Sporadic campaigns had been held in the pre-Confederation years. But the opening of Western Canada for additional settlement in the late 1880s spurred the concept to new heights. The various land agencies, both public and private, had millions of acres of land to sell. The government of Canada wanted settlers at any cost. Private companies such as the CPR, Hudson's Bay Company and various land companies wanted to unload their acres for profit. The Hudson's Bay Company had acquired its acreage through the sale of the Northwest Territories in 1869. The Canadian Pacific Railroad was given millions of acres to finance the transcontinental railroad. And various companies had purchased or acquired Western land for speculation. Large-scale settlement was impractical before completion of the Canadian Pacific Railroad in 1885. But after that date, bring on the settlers! There were several methods employed to attract new immigrants. Land agencies, advertising campaigns, brochures, newsreels, sightseeing trips (all expenses paid) and glowing testimonials were all utilized.

Canadian foodstuffs exhibit, St. Louis Exposition, 1903. An impressive array of Canadian food products interspersed with photographs of their places of origin. Wheat and flour are heavily emphasized, but canned goods are also prominent. Such displays at major exhibitions around the world had two motives: to sell Canadian goods abroad and attract immigrants. They were successful on both counts.

Displays at the international trade shows, fairs and exhibitions portrayed Canada as the "Land of Opportunity." To this effect, huge displays of agricultural products were often showed. After all, it was farmers that were wanted. Other items such as flora, fauna and landscapes were also used to promote the romantic image of Canada, and especially Western Canada. There were no pictures of snow-covered expanses or desolate frozen wastes, no panoramas of dense bush, and of course, no blackflies. No tales of failure or hardship were published. Prospective immigrants were shown only Canada's best side.

The advertising world has changed radically in recent years: mass-circulation newspapers and magazines, television, other electronic media and the Internet are all now used. The annual fair is certainly no longer the only game in town. However, fairs have not lost their commercial element, and they are still a popular venue for businesses and manufacturers because they still attract large crowds in a friendly, genial, one-on-one atmosphere. There is something about the feel of a fair that still attracts both sellers and buyers in large numbers.

But let's have the photographs do the talking. Welcome to the commercial side of fairs:

Kent County exhibit, CNE, 1928. Sunny Kent County in Ontario advertises its finest agricultural products.

Hastings County, Ontario, exhibit, CNE, 1920. A curious mix of dairy products, honey, apples, grain and, tourism. Fishing and hunting are promoted in this "the ideal stomping grounds for tourists."

Swift's Meat Packers display. A whole line of specialty products are featured here: processed meats, ham, Carnation milk, pickles, soda biscuits, Crisco, HP Sauce and ketchup are just a few featured items. The cooked meats range from 34¢ to 85¢ per pound, while butter sells for 29¢ per pound. The whole setup is called "provisions," a term whose meaning has changed over the years. Many of these items and brand names are still with us a hundred years later. However, the pricing and packaging have changed, slightly.

South Qu'Appelle District, Saskatchewan, display at a fair, circa 1900. Such displays were clearly intended to impress prospective farm settlers to the region. Note that there is both a man and a woman in the photo. The male representative would sell Canada to the husbands, while the woman rep would talk up Canada to the wife.

An impressive display promoting agriculture and rural living at the Regina Fair, 1923. Government exhibits such as this were designed to educate the farmer and imspire him to improve his condition. They were also "feel-good" displays, reminding rural folk of good times past — even when conditions at the time were tough in the farm industry.

Displays by the Manitoba Department of Agriculture and the Manitoba Agricultural College at Brandon Fair, 1930. Government at all levels have always been big participants at fairs. Government recognized the value of agricultural fairs as educational mediums, and so supported fairs with both grants and displays.

A display of fruit from the Okanagan Valley, British Columbia. This display was designed to promote B.C. apples outside the province. Many of the apples are carefully displayed in shipping cartons to give the prospective buyer a good look at the product. Apple packing was a skill, and many fairs held apple-packing competitions.

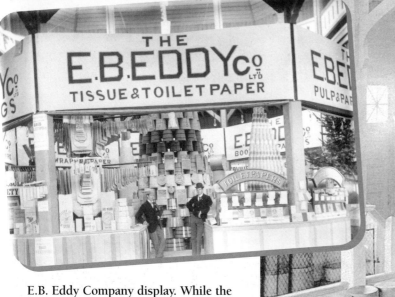

E.B. Eddy Company display. While the famous Canadian paper company sold paper bags, wrapping paper, washtubs and that old favourite, the scrub board, its pride and joy was that new-fangled invention: toilet paper. If you have a great idea to sell, take it to … the fair.

Cheese display, CNE. In the two decades before 1900, cheese was Canada's number-one agricultural export. Canada east of Manitoba was dotted with cheese factories and dairy farms. Such companies as this were only too willing to show off their products at the fair.

Exhibit of James Ramsey, hardware merchant, Edmonton Fair, 1916. Ramsey was certainly aiming at the farm market. In the photo are cream separators, butter churns, sewing machines, wringer washing machines and cook stoves: the cutting edge of modern technology, at least in 1916! By the way, the horse was either stuffed or a really, really good actor.

Canada Post exhibit, 1923. When Canada Post wanted to promote its services, it did the logical thing and joined the fair circuit. Notice the emphasis on international or overseas services. Useful tips are provided on everything, from where to place your stamp and how to wrap packages, to how to handle fragile items.

Imperail Oil Company display, 1910. Gasoline was still handled in drums because gas stations were few and far between. In many rural areas, car owners had their gasoline shipped in by train and transported home by that old standby, horse and wagon. It would take a few years for the auto-service industry to catch up to the spread of motor cars.

Drewry's Brewery display, 1909. This bottler handled both beer and water. Besides attractive displays of product, the vendor tosses in a healthy dose of Canadiana (a model railway, pictures of local Canadian landmarks and even photographs of King Edward VII and Queen Alexandra). Wow! How much more Canadian can you get?

Above: Manufacturers' displays, 1913. Two marvellous new inventions, flush toilets and vacuum cleaners, are exhibited. The flush toilet was called a "water closet." Its design hasn't changed a whole lot since then. Fairs were good opportunities for manufacturers to spread the word about their products. Most Canadians "saw it first" at a fair.

Above: A commercial display selling kitchen supplies. The emphasis appears to be on hotel kitchen equipment, but considering the size of many families a hundred years ago, many women must have felt as if they were running a hotel kitchen. While some of the gadgets displayed are outdated by today's standards, others would still look good in a modern hotel.

Machinery row, Saskatoon Fair, 1950s. Tractor-drawn equipment has replaced the horse-drawn variety, but machinery row is still the largest mall for farm-equipment sales in the community. This display caters to more than just farmers, as evidenced by the road graders in the foreground.

Concession area, Fredericton Fair, New Brunswick. On the right is the two-storey exhibit hall. The ever-present tent city is on the other side of the midway. And in the foreground, that old reliable page-wire fence (great for leaning on). Also present in the foreground are a dunk tank and a crown and anchor wheel, both long-time favourites at any fair.

Concession area, Brampton Fair, Ontario, 1924. On the right is a display of Ford tractors. It was clearly a late-fall fair, as everyone is heavily bundled up against the cold. Hats are clearly in fashion for both men and women.

Gas-appliance display, 1910. For urbanites, gas-fired appliances were the apex of technology. For rural dwellers, it was back to the woodpile (or coal scuttle). Notice the ovens on top of the stoves.

WORLD WAR I

Canada was plunged into a world war for the first time on August 4, 1914. Canadians were naïve in 1914: they had no idea of the scope of "total war." The nation was totally unprepared and blissfully ignorant of what was necessary to mount such a war effort. The first response was to round up a contingent of volunteers and militia, the so-called weekend warriors, and ferry them to Europe as quickly as possible. After all, everyone expected the war to end by Christmas 1914! When it became obvious the war was going to drag on, more long-term planning became necessary. One of the most obvious needs was securing facilities for training soldiers. Many towns across Canada had armouries or drill sheds, but permanent military bases were sorely lacking. Consequently, military planners cast their eyes on the best chain of facilities available: the fairgrounds. Larger fairgrounds contained wide, open spaces for drilling and manoeuvres, and in many cases large buildings to house and train soldiers.

Beginning in 1915, strategically located fairgrounds were taken over by the military to house, train and drill the new mass armies. Agricultural societies were proud of their role in the war effort. But the military occupation did cause problems. Some fairs were cancelled outright while others were downsized, and some moved to outdoor-only fairs. In other cases, the army moved out of the fairgrounds for the week of the fair. For example, the CNE grounds was occupied by the military but the exhibition went ahead as scheduled. The army simply moved to summer quarters

Machinery Hall, CNE, 1916. Banfield and Sons rented the building to make shells for artillery guns, eighteen-pounders to be exact. Posters placed around the building exhort workers to toil "For King and Country." It took a few years, but eventually Canada was prepared for "total war."

A regiment in training, Vancouver Fairgrounds, World War I. The local fairgrounds were an ideal site for army training, both temporary and permanent.

Trench display, Peterborough Fair, Ontario, 1917. While the Great War raged in Europe, these mock-ups were used at fairs to aid recruiting and raise funds for the war effort. The grim reality of trench warfare was not displayed.

at Niagara-on-the-Lake, thus freeing up the fairgrounds for the annual exhibition.

Obviously, the war permeated the fairs in many other ways. A few societies cancelled their 1914 fairs out of respect for the war effort. But cancelled fairs were of no benefit to the men and women fighting overseas, so 1915 included a full schedule of exhibitions once again. When the Canadian government asked farmers to drastically increase all foodstuff production, field-crop competitions at fairs were strengthened. By 1915, Ontario fairs had swelled the number of field-crop competitions to 258 from only 159 in 1913. Military recruiters also concentrated at the local fair, the obvious place to reach both urban and rural audiences. And fundraising was everywhere, headed by all branches of the war effort, but especially the Red Cross.

Horses had been prominent at the fair but, during the war, they became even more important. Brandon fairgrounds, for instance, became a cavalry training depot, appropriate for the horse capital of Western Canada. Western fairgrounds in London, Ontario, were the scene of more frequent horse sales. It was estimated that the Canadian army purchased nearly 50,000 horses for army service from this section of Ontario alone. Many of the horses were stabled and trained on those fairgrounds.

CANADA AND WORLD WAR II

In 1939, Canada again found itself embroiled in a world war. But this time, Canada was less naïve, and the lessons of "total war" in the last conflagration were not forgotten. The nation geared itself for a long war: no more haphazard arrangments like World War I. The Department of Defence immediately designated which fairgrounds were to be used as training depots. Dozens of fairs were cancelled outright as the army moved in for a long occupation. Thirty-three Ontario fairs were cancelled for this reason in 1940, and other fairs operated on a reduced basis. Often, the army would occupy the fairgrounds for fifty weeks of the year, moving out in time for the local fair, while other fairgrounds were only partially occupied by the military. By 1942, Ottawa, Regina, Peterborough, London, Saskatoon, Edmonton, Calgary, Toronto and Vancouver fairgrounds, to name just a few, had become military camps, primarily training depots for new recruits. Other fairgrounds had more specialized uses, such as Medicine Hat, which became Service Flying Training School No. 34. The Lethbridge fairgrounds housed a prisoner-of-war camp for captured Germans. One night, eighteen of these prisoners tunnelled their way out of camp and disappeared. They must have made it over the border to Montana, for they were never recaptured.

Fairs were also prime grounds for recruiters. No wartime fair was complete without a military recruitment booth. The Royal Canadian Air Force signed up ninety-six recruits at the 1941 Saskatoon Fair. Simcoe (Ontario) Fair, as a sign of support for servicemen, fed and entertained trainees stationed at their fairgrounds in a special "Appreciation Day." That day, over one thousand servicemen enjoyed a one-day holiday. Fairs were also very active in fundraising. The T. Eaton Company sponsored the "Hogs for Britain" program, which gave $30 to each fair that sponsored a pen of five bacon hogs, which would then all be shipped to wartorn Britain. Forty-five Ontario fairs held this class in its first year. Collingwood (Ontario) Fair donated all entries in its apple classes to the war effort. Teeswater Fair in Ontario was one of many fairs to charge a fee to view "Resolution Moneybags," a 3,200-pound bull, with proceeds going toward the war effort. Other fairs just gave cash, such as the one in Lindsay, Ontario, which donated $1,500.

A common debate during the war was

Parade of champions, Armstrong Fair, B.C., 1940s. To do their part for the war effort, exhibitors lined up animals in a huge "V for Victory" pattern. A white chalk line marked the spot for animals and exhibitors alike. It took a bit of persistence to get most animals in formation: a few couldn't remain still long enough. During the early 1940s, the war effort was on everybody's mind, and was a significant part of fairs coast-to-coast.

whether fairs should actually be held at all. Many Canadians believed all fairs should be cancelled and the money and time poured into the war effort. But others believed fairs had an important role to play on the home front. They not only encouraged increased agricultural production when it was needed most, but also provided an important diversion for workers and military men and women at a time when many entertainments were not readily available. Morale on the home front was an important consideration in keeping the fair circuit going.

But not all fairs could keep going during the war. Gasoline rationing and restrictions on railway travel took a toll on the fair industry and on society in general. Some winter fairs, which required a lot of hauling, were cancelled. Midways were hampered by restrictions on travel and American midway shows and carnivals were banned from crossing the border. Patty Conklin had his midway show halved by such restrictions. Originally, his midway was denied permission to use any railways, but this meant many fairs would be without a midway. The fair organizations gathered their support and lobbied on Conklin's behalf. In true Canadian fashion, a compromise was reached whereby Conklin's midway was allowed one-half its pre-war railway transport. The show must go on! At the end of the war, the military slowly evacuated its occupation of fairgrounds. It took a couple of years for the fair circuits to get back to normal.

A fairgrounds converted to military use during World War II. The before-and-after shots of the Peterborough, Ontario, fairgrounds. Barrack huts fill the infield, while the permanent structures in the background were used for training, mess halls, etc. No wonder the Peterborough fair was cancelled during the war.

The Exhibit Hall: Pumpkins to Pictures

EVERY FAIR HAS ONE EXHIBIT HALL, and to many they are the backbone of the agricultural fair. The exhibit hall is a term loosely used to describe the building where various and sundry items are all on display. These items can vary dramatically, but at most fairs they consist primarily of sewing, baking, crafts, school or junior work, photography, art, vegetables and flowers. These are entered by individuals in competition for prizes. After judging, they remain on display for the public to view. Most fairs have hundreds of different classes open for competition. People enter for many reasons, but the primary goal is pride in handiwork, be it a quilt or a pumpkin. It is an opportunity to show off, or be rewarded for one's artistic, culinary, horticultural or homemaking skills. The exhibit hall entry lists were once classified under "domestic work" or "ladies' work," but today they are called homecrafts at most fairs.

In the Beginning

Early European fairs did not have exhibit halls where handicrafts were entered for prizes. However, the earliest North American fairs offered prizes for vegetables and grains, and by the early 1800s, whole sections of domestic or women's work had been added to the fair program. Elkanah Watson, founder of the Berkshire fair system, championed the addition of these categories. The concept took hold, and soon every fair had a domestic arts section.

The exhibit hall often reflected the times. In pioneer days, practical classes such as butter making, rug hooking, vinegar making and hat making, to name a few, were very common. As times changed, prize lists changed to reflect changing attitudes, fashions and technologies.

There were two main objectives behind domestic arts exhibits: it was a chance for the housewife to show off her skills, be they culinary or artistic; and the exhibit provided educational opportunities for those involved in the competitions. By observing prize winners, the women

Exhibit Hall, Nelson Fair, British Columbia, 1905. An attractive exhibit hall with a liberal use of flags, including the Union Jack, the B.C. flag of the day, and even the French tricolor. The tables are packed with entries while displays flank the wings.

could learn ways to improve their skills. As one prairie woman observed:

"Even if we didn't get any prizes for our cooking at the fair, at least we did compare critically our own efforts with those of others, see where we fell down and resolve not to make the same mistakes again. In this way we advanced a distinct step for the rest of our lives."[1]

Accordingly, another contestant observed about her fruit cakes:

"They did not measure up with the prize winners. I've wondered why, for I had taken all manner of pains with them. And the reason? Why, I had neglected to clean out the range under the oven, and it was choked up with soot."[2]

Ironically, the above problem was partially responsible for a first place in the fruit-cake class. The diffused heat prevented the fruitcake from burning. The lesson learned:

"Henceforth for fruit cakes, while I'll have the space under the oven clean, I will place an inverted pie plate under each cake to get the same effect."[3]

Wow! Practical education at its best.

Judges and Judging

Homecraft judges have had their problems over the years. Disputes with exhibitors were common, many going all the way to the top. One such case involved Mrs. Johnson, a self-styled "culinary expert" with a resumé of red ribbons from many Alberta fairs. At the Ranfurly (Alberta) Fair, she entered three cookie classes and netted two seconds and a third. This was a damaging blow to both reputation and ego, especially when the prize winners were considered "inferior in quality." Her cookies were "dainty and crisp" while the winners were "clumsy, bigger than the palm of my hand." The distraught Mrs. Johnson concluded her letter with "My interest in fairs is killed!" The task of replying to this letter of complaint was given to the Superintendent of Alberta Fairs, who replied that the judge, although trained in culinary arts, was a rookie in her first year of judging. He admonished the "veteran exhibitor" with these words of advice:

"I know it is provoking in the extreme not to get prizes if you are entitled to them ... but every exhibitor of any note has had this experience ... on the other hand, those who exhibited a lot often got prizes they did not deserve. Over a lifetime, such exhibitors usually break about even."[4]

For anyone who has ever wanted to exhibit at a fair, these are words of wisdom.

Exhibit hall, Prince Rupert Fair, B.C. Displays in the background include Nabob Coffee, McRae Brothers, Pattison & Ling Appliances (motto: lightening the labour of the home) and Ormes Ltd. (pioneer druggists). What is a "pioneer druggist" and how do they differ from regular druggists? Inquiring minds want to know!

School fair or Junior exhibit, Renfrew County Fair, Ontario. An exhibit tent was devoted to entries from children. Everything from grain to flowers to vegetables to baking was, and still is, included in this competition. Junior exhibits have always been a big part of fairs in Canada.

Pie Classes.

To be sure, strange things have occurred in the exhibit hall. Mrs. Jean Muldrew, for instance, was a well-respected homecraft judge, and a keen observer of proper judging etiquette. When she asked to taste the preserves at one fair, the astonished director replied that, well, *we've never tasted the entries before!* Mrs. Muldrew changed that practice at once and soon the most beautiful jar entered was found packed in pure acid and totally unfit for human consumption. Not only was the jar disqualified, but it was learned that the same entry had won several years in a row! Another favourite trick was to pack raw rather than cooked vegetables and fruits in sealers. They looked great but spoiled immediately upon opening. Mrs. Muldrew also sliced every loaf in half, another novel idea to some directors. When criticized by a director for disqualifying a particularly good-looking loaf of bread, the judge urged the critic to smell the loaf. After vomiting, the director agreed with the decision and let the judge do her job. Another irate exhibitor tried to prevent the judge from cutting her cake. She blurted out that she planned to show it again next week at another fair. After all, the same cake had already won four first prizes that year![5]

In the sewing category, many complained that some handiworks were shown year after year. The standard rule stating that "all exhibits must be made in that past year" was difficult to enforce. One widower at an Ontario fair was accused of keeping his late wife's memory alive by showing her favourite quilt year after year. After eleven consecutive first prizes, the quilt fell apart from age. Another age-old problem in the exhibit hall is the handling of exhibits. This often soils or damages the them, especially those in the sewing and crafts exhibits. To protect these valuable goods, many fairs began to enclose certain sections behind fences or barriers to keep spectators out. Certain exhibits have also been subject to theft, especially at the close of a fair. So many fairs only allow proven exhibitors inside exhibit halls at closing time. One novice exhibitor who lost her peach pie under these circumstances became philosophical about the whole thing. She was sad to lose her peach pie, but happy that someone thought it worthy of the crime. Another common complaint focused on storage. The Portage-la-Prairie Fair for instance, featured sixty entries in the butter class. The butter was on display for the entire length of the fair. In the days before refrigeration, this was a problem. By fair's end, the smell of rancid butter overpowered the odours from the livestock barns.

Men and Women at the Exhibit Hall

There was a shortage of women in the days of early European settlement of the Prairies. This translated into a lot of single men "batching" on the homestead. Such inequality among the sexes also filtered through to the local fair. One bachelor, W.A. Trounce of Saskatoon, sponsored the following classes at the 1887 Saskatoon Fair, open to bachelors only:

1. Best darned socks 50¢
2. Best patched pants 50¢
3. Best two pound loaf of bread

It's not on record if these classes were designed to end the state of bachelorhood by attracting potential mates, or simply a reflection of the daily state of bachelor affairs.

Female exhibitors, on the other hand, were a little more restricted within their traditional roles. While women's work was a part of every fair, it did not always enjoy equal status with men's work. The earliest fairs placed much more emphasis on livestock, horses and grain, to the detriment of domestic crafts. One male director spoke for many when he proclaimed: "The spirit among the men has been that we must get the very best judge there is to judge cattle, sheep, swine and horses. But anyone around here can judge women's work."[6] Ouch! It was not pretty to watch a ham-handed cowhandler judge the crochet class.

Craft display, Grand Prairie Fair, Alberta, 1913. The Peace River District was one of the last frontiers in Alberta. As settlement advanced, so did fairs. The inside of this exhibit hall has a grass floor, so it must have been a temporary structure.

Another male attitude that angered women exhibitors was expressed by a frustrated exhibitor:

"Fancy work, whether appliqued, Richelieud or embroidered is discovered to be his idea of *vanatis, vanitatum* ... the Lady exhibitor has

Exhibit Hall, Priddis and Millerville Fair, Alberta, 1912. A variety of exhibits, including flowers, canning, vegetables and grain are seen in this photo. Exhibit halls like this were very common at small fairs all across Canada.

Ukrainian homecraft display at Melfort Fair, Saskatchewan. One reason fairs were considered worthy of government grants was the role they played as an assimilation or inclusionary tool for new immigrants. People from all backgrounds gathered or participated at the local fair. The exchange of ideas and knowledge made fairs ideal educational tools for the growth of a unique Canadian culture.

only to put her odd minutes work into her pocket and march off with a sum out of all proportion to the time spent or benefits given."[7]

Clearly, the gentleman espousing this theory had never quilted in his life. Often the lady spent long hours producing fancy work to be shown for prizes of as little as 25¢. While vanity certainly played some small part in the domestic arts division, the so-called men's classes were not above reproach. As one insightful woman pointed out, the same educators who criticized fancy work as vanity, immerse themselves in horse racing. The kettle just called the pot black!

Women directors actively worked hard at improving their division. Practical classes were encouraged over functionless exhibits. For example, rather than encourage cushion covers so elaborate that "mere man is warned never to put his head on them," domestic improvement was emphasized. One important branch of domestic improvement involved the constant cavalcade of labour-saving devices presented at fairs. These ranged from new wood stoves to apple peelers to recipes to methods of clothing manufacture. While fairs were not in the business of invention or manufacture, they did serve as showplaces for the new and improved. Eager fair-goers crowded the exhibit hall and commercial exhibits to see the latest inventions. Fairs were an important link between industry and the household. While the menfolk cruised machinery row or gawked at the newest automobile line, the womenfolk eagerly scanned the displays for everything from new food products to kitchen gadgets to the newest fashion.

At many early fairs, married women did not place exhibits in their own names. They often showed under their husband's name. At the first Moose Jaw Fair, John Lawson, for instance, was credited with winning four first prizes for the following:

best calf	best ox-team race
best home-made bread	best baby daughter

Agricultural societies and their fairs mirrored the society of the era. Women were often discouraged from becoming directors and participating in decision making. Many men were of the opinion that women belonged in the exhibit hall. The earliest fairs featured no women on the board of directors. However, by the turn of the twentieth century, women began to pursue more lofty positions on fair boards, and in society in gen-

Butter-making competition, London Fair, Ontario. It seems to be a "women only" affair. All competitors used the same equipment. Clearly, the show was a fan favourite, as evidenced by the full gallery.

eral. By 1930, it was estimated between 5 and 10 percent of fair directors were women: still puny numbers, but a trend was clear. By the end of the twentieth century, almost all fairs had women directors on their boards. The name "women's division" was replaced by "homecraft division" to mirror changing attitudes. And gender became less and less of an issue at the fair.

The Professional

A problem at many fairs was the presence of so-called "professional" exhibitors. Many amateur exhibitors felt they did not stand a chance when competing against them. This begs the question, "What is the definition of a professional?" Since that question has never been answered fully, fairs cast about for solutions to keep the dreaded professional, whatever it may be, from their shows. The most common method was to limit entries to locals. After all, it was easier to tag the outsiders as "professionals." Entrants who did not have a local address were often regarded suspiciously. On the other side of the coin, it was argued that fairs should be rewarding the best exhibits and where they came from was unimportant. Still, many fairs across Canada closed their prize lists in various ways. In response, government agencies usually looked the other way in domestic-science classes, but when the livestock

Exhibit Hall, Prince Rupert Fair, British Columbia. Another attractive display of vegetables, flowers and more commercial exhibits, including one by an engineering company and a photographer. Multiple usage for such a hall is evident by the basketball hoop on the left.

Fruit exhibit, Horticultural Building, CNE. This was not only a competition but also a chance to market some products. Displays from major fruit growers can be seen in this picture. The potted palms add an exotic touch.

entries were restricted to locals only, they stepped in. Many provincial governments withheld grants for societies that closed their prize lists in this way. Other fairs proudly announced their prize lists were "open to the world."

Grain and Field Crops

Fairs also include vegetable, flower and grain classes in their exhibit halls. While not exactly the outcome of "domestic" science, these products of the land have naturally found themselves partnered alongside others inside the exhibit hall.

At early Canadian fairs, grain-and field-crop competitions were very important parts of the agricultural component of the fair. Almost every farmer raised some type of grain. Improvement of the grain/field crop industry was a prime goal of agricultural societies. To work toward this goal, special seed fairs were often held in the spring, allowing farmers to examine new varieties, compare the merits and drawbacks and perhaps, purchase the product. The fall fair was an excellent stage for presenting the results.

In the latter half of the twentieth century, the total number of grain farmers declined, and farming became much more specialized. Grain farming became regionalized, with wheat farming centralized in the West. As a result, the grain- and field-crop competitions have virtually disappeared from many fairs, with each region of the country adapting to local specialties. In Western Canada, wheat remains the big field crop, while in Ontario field corn has become important. Fairs have adapted to reflect these regional differences.

As with any other type of exhibit seen at fairs, the field-crop division is not without its moments. One clever housewife placed her grain in a butter churn lined with flannel and gently polished it. The resulting glossy entry impressed the judges into awarding it first prize. The very first Brandon fair featured a poor turnout in the grain classes. One resourceful director went to the local grain elevator and "borrowed" enough samples to beef up the competition. Turnips were an important crop on many early farms. A few acres of this root crop supplied excellent winter feed for livestock. Every turn-of-the-century farm had a turnip chopper, or cutter, among its machinery. The night before the local fair, one Orono, Ontario, farmer noticed a light moving in his turnip patch. Closer examination revealed his neighbour gathering some specimens for the next day's fair. As fate would have it, the turnip grower captured first prize, while his cheating neighbour placed second. The neighbour then had the gall to question the turnip grower on how he had beaten him. The farmer acidly explained, that he had picked his roots by daylight, unlike the thief.

Vegetable and flower judges also have tales to tell. One real bugbear faced by judges over the years was

Exhibitors pose proudly with their prize winners outside the hall, Grand Prairie, Alberta. All ages and both sexes have brought their best to be judged.

the question of what comprises a prizewinning entry. Some exhibitors, and judges deem the largest entry to be first-prize material. In some classes this works, but in others it doesn't. Pumpkins and squash are easily judged for size. But for anyone who has attempted to process a monster turnip or huge potato, size does not mean perfection. For many garden vegetables, oversize means of course, rough and inferior. Therefore, judges were encouraged to reward moderately sized but perfectly formed vegetables. Eventually, many fairs began to separate their vegetable classes into table-size fruits/vegetables and the huge, monstrosities. In recent years this latter section has grown in popularity. Giant-pumpkin classes have become highlights at many fairs. Some gardeners go to great lengths to

Display of garden produce, Edmonton, early 1920s. A newspaper article about the fair reported carrots on display that measured two feet long. The length was carefully certified so that Calgarians could not dismiss the behemoths; after all, the prize winners at the Calgary Fair were only 18 inches long. Take that, Calgary! Just another chapter in the friendly rivalry.

A 403-pound pumpkin grown by William Warnock of Goderich, Ontario. This pumpkin won first prize at the St. Louis World's Fair in 1904. The world record for giant pumpkins, as of the year 2004, was 1,385 pounds. But as you can see, growing "the big one" was also a popular pursuit a hundred years ago.

produce these monoliths. The World Pumpkin Contest held annually in Owen Sound, Ontario, attracts serious pumpkin growers from all across Canada. There are many tricks to growing the largest pumpkin, and some growers even put up a tent over the potential champions to keep the weather from interfering with their march towards the top prize and a world's record. The current world's record stands at 1,018 pounds.

Horticultural societies and flower shows have traditionally been associated with fairs. Almost every fair in Canada has flower classes in their exhibit halls. Many communities across Canada also have horticultural societies that are closely related and administered by the same branches of government. While many horticultural societies hold flower shows separate from those at the fair, others participate at the local fair. Early in their history, agricultural societies were designated as umbrella organizations under which other groups were formed.

Arts and Crafts

Throughout their history, Canadian fairs have included arts and crafts classes in their exhibit halls. Painting, handicrafts, woodworking and photography are just a few. The sections change with the times, much as fairs themselves do. Exhibit halls also reserved sections for junior exhibits and school work. The "school fair" is a big part of most fairs across Canada. When separate school fairs dissolved in the 1940s, the agricultural societies stepped in to fill the void. Encouraging young people to show at the fair has always been an important part of agricultural fairs.

The domestic or home craft division has changed dramatically over the last 200 years. Many, many styles and trends have come and gone in the domestic sciences division. Yet each year, exhibit halls all across the country are filled with thousands of individual articles, lovingly prepared to win a prize at the local fair. Competing at the fair makes the whole fair experience more meaningful for many fair-goers. The exhibit hall is one part of the fair where many show off their accomplishments.

Floral Hall, CNE. The large number of entries reflects the popularity of horticulture in urban centres across Canada at the turn of the century. Competition in the horticultural classes was intense. Aster exhibits grace the foreground, while roses decorate the next table.

CRYSTAL PALACES: THE FAIR CATHEDRALS

One of the most common features at fairs in the late 1800s was a "crystal palace" exhibit hall. The first great world's fair in London, England (1851), was dubbed the "Crystal Palace Exhibition" because the outstanding feature was the massive and magnificent Crystal Palace exhibit hall. This engineering marvel was 1,800 feet long and 465 feet wide. It had two floors and was framed almost entirely by glass panes, some 30,000 of them. It contained exhibits from all over the world. Canada, the largest British colony, had 120 exhibitors presenting their goods. The glass walls were designed to utilize the maximum daylight and add a bright, cheery, modern atmosphere to this showcase of industry. The Crystal Palace was a structural success and an aesthetic marvel. Visitors raved about the design and the beauty of the hall. As a result, copycats sprang up all over the world. Crystal palaces were built at fairs everywhere, the larger and richer fairs building larger and more elaborate replicas of the glass hall. The smaller, more humble fairs built more modest versions, built to the scale they could afford. The bigger fairs used glass and added magnificent cupolas and domes. The smaller ones used wood and a few windows to try to capture the same effect on a more modest budget. Many halls were named

Crystal Palace, Picton Fair, Ontario. A modest Crystal Palace at a B fair. Notice the cross shape and single cupola. The elaborateness of the Crystal Palace reflected the size and affluence of the fair. This building has been restored and is still in use today.

Crystal Palace, Central Canadian Exhibition, Ottawa. A less elaborate crystal palace, but heavily decorated with cupolas. The structure has served multiple purposes over its career, and has been home to manufacturers displays, a hockey arena, and a "cattle castle." It is still in use today.

The second Crystal Palace, CNE, 1884. The original Crystal Palace was constructed for the first Canadian National Exhibition in 1879, but had its career cut short by fire. Notice the elaborate outside facades and the extensive use of glass. The glass windows were designed to capture the maximum daylight and create a bright affect within the palace.

"Crystal Palaces" when they maintained no resemblance to the original structure at all. They simply wanted to capture the magic that term evoked in the fair industry. Most of the crystal palaces have been lost over time, fire being the main culprit. Many a palace just fell victim to age and neglect. Only a handful exist today, monuments to a bygone era when the term "crystal palace" evoked memories of style and opulence.

Crystal Palace, Calgary, Dominion Exhibition, 1908. Hosting a major exhibition, such as the Dominion Exhibition, meant fairs were funded to construct large and visually impressive exhibit buildings. This fine example of a crystal palace housed industrial displays. Notice the formal garden in the foreground and the extensive use of glass windows to diffuse light inside the building.

SEED FAIRS

One original goal of early agricultural societies was to provide farmers with improved livestock and seed varieties. To the latter end, many societies sponsored special seed fairs. These one-day events allowed interested farmers and businesses to buy and sell seed grain. They were usually held in the early spring just before sowing time and were often sponsored by the local agricultural society.

Western Canada was a hotbed of grain growing, and by extension, seed fairs. Every group advertising for settlers on the Prairies included a liberal sample of prize-winning grain in its exhibit. The Canadian Pacific Railroad would often buy prize-winning grain entries and include them in their advertising displays. Western Canadian wheat began to win championships at international grain fairs very early in its agricultural history. The famous Red Fife Wheat from the Prairies won at Antwerp, Belgium, as early as 1885. When Canadian Prairie wheat captured top honours at the Columbian Exposition at Chicago in 1893, and the Pan-American Exposition of 1901, the world began to take notice. Seeger Wheeler started an amazing string of grand championships for prairie wheat in 1911 when he captured grand honours at the World Grain Exhibition. Wheeler captured five championships in eight years, losing three times to other Canadian Prairie farmers. Between 1911 and 1955, Prairie wheat growers won an incredible 35 championships. Herman Trelle of Stavely, Alberta, matched Wheeler's record by also capturing five world championships.

In the two decades between 1905 and 1926, Saskatchewan alone held 894 seed fairs, an average of 44 seed fairs a year. In 1933, at the height of the Great Depression, Regina hosted the World Grain Show. Scores of countries from around the world sent their best grain to be judged and, once again, Canada showed its dominance. Herman Trelle defeated 95 other entries to win for Red Spring Wheat (soft wheat). Trelle vanquished 64 other competitors to win the hard-wheat class. First prize was a thousand dollars and a boatload of prestige, both for the grower and the nation as a whole. Agricultural societies and seed fairs had proved their worth.

Livestock: Four Feet and Feathers

A fair without animals is like a kite without a tail: it just doesn't go anywhere."[1]
—a Weyburn, Saskatchewan, farmer, 1952

No true agricultural fair is complete without livestock. The earliest agricultural societies believed livestock improvement was one of their primary reasons for existing. Such societies used two methods for improving livestock. Many societies imported purebred livestock and leased or sold the animals to prospective farmers, but profit was not the primary goal. They were solely interested in improving the quality of livestock in their area. They acted as importers and cooperatives for the livestock industry. Their role was vital to a pioneer farm economy. The pioneer farmer could not afford to purchase purebred livestock: they often lacked capital. In many areas, the livestock was not even available until the agricultural society imported it. But as the farm sector matured across Canada, agricultural societies abandoned the livestock cooperative system and moved into their second method of promoting livestock improvement: the show ring at the fair. This shift occurred all across Canada.

Poster promoting ideal cattle breeds, 1910. Representatives from the Ontario Ministry of Agriculture assembled this collage of prizewinning cattle at Ontario fairs. The poster was hung in agricultural offices across the province.

The Show Ring

The show ring was, and still is, the ultimate test for the livestock breeding industry in Canada. Agricultural societies reasoned that by offering cash prizes, show facilities and annual dates, they could attract livestock breeders and buyers to their fairs. They were bang-on correct. It became mandatory for any farmer buying or selling breeding stock to attend at least some fairs every year. There animals were shown off in public, ideas were exchanged, sales arranged and networking done. Fairs provided excellent exposure for the breeders. The show ring became the place for breeders to evaluate their efforts. A first-place ribbon at one of the elite fairs such as the Royal Winter Fair, Toronto, Agribition, Regina, Brandon Fair or the Calgary Stampede was a huge boost to the value of any animal. To this end, livestock exhibitors developed an elaborate series of techniques, styles and methods of showing animals.

Parade of Champions, CNE, late 1890s. The Hereford winners are parading before the grandstand. In the background, other cattle-class champions await their moment of glory in front of the massed patrons. This Parade of Champions was a very common sight at fairs throughout Canada.

Showing livestock at a fair was not the easiest job in the world. In the days before motorized transport, the first step was an often long and arduous trek from the farm to the fairgrounds. Small animals such as pigs, sheep and chickens could be brought by wagon; but for the larger beasts, it was one plodding step at a time. A good speed was three miles per hour, so it would often take several hours to arrive at the fair. The animals often protested the move from familiar and comfortable surroundings and the forced march to the unfamiliar fairgrounds. Having to play cowboy first and see the fair afterwards also cut into the enjoyment of it all – especially for the farm children. And even worse than the long walk to the fair was the seemingly longer trek home! Many a young farm girl or boy must have felt discouraged as they trudged home with their weary livestock while sans-livestock friends zipped by in buggies and later, motor cars. It must have been enough to make them want to give up the whole fair game.

The arrival of the railway and later, motorized trucks, certainly changed livestock shows. Speed increased more than tenfold. Distances shrank. It was now possible to truck animals long distances to a variety of fairs — not just the local one. Farmers developed circuits, often attending at least one fair every week. Many exhibitors began to travel outside their area, and indeed all over Canada. Livestock exhibitors used

Grandstand, Sherbrooke Fair, Quebec, 1912. The huge grandstand is filled to capacity for this event. For those who think livestock shows do not attract crowds, think again!

Canada's fine railway network to attend major shows like the Royal Winter Fair, Brandon Fair, the Calgary Stampede and the Atlantic Winter Fair in search of more exposure for their animals. But the railroad had its disadvantages. After their arrival at the railway yards, the exhibitors and their animals were forced to parade to the fairgrounds. And fairs without railway access were left out of the circuit altogether. With the subsequent increase in exhibitors, the "better" fairs flourished, and fairs that did not have good reputations among exhibitors had difficulty attracting decent entries.

This change had some drastic effects on some of the smaller fairs. Many disappeared over time because they could not attract sufficient exhibitors or fair-goers who preferred attending the larger or better fairs. Livestock exhibitors became more professional and fewer in number. The big professional breeders with their purebred, registered herds began to replace local farmers. The professional breeder wanted a circuit of larger fairs that offered better prizes and more exposure. In order to promote their herds, some top breeders would rent a livestock rail car and spend weeks on the road, travelling from fair to fair all across Canada and the U.S. It was a transient life, taking herds from place to place, riding and sleeping on railway cars. One circuit in Western Canada started in late June and ran chronologically through Brandon,

Calgary, Edmonton, Saskatoon, Regina, Great Falls, North Dakota, Billings, Montana, Vancouver and Victoria. It required a total of ten weeks on the road. Many exhibitors confessed that they made little or no profit from the prize money on this circuit, but they did get good publicity and free feed.

The "professional exhibitor" became an issue at some fairs. Some professionals purchased a complete set of prize-winning animals and toured the fairs for profit. They contributed little to the agricultural industry. In the poultry business, these professional exhibitors were called "string men." A complaint from the Carmen Fair, Manitoba, charged that three railway cars of poultry attended their fair, gobbled up the prize money and moved on to the next show to repeat the same performance. Most of the birds were exotic breeds of no practical use and were simply nice to look at. It was also duly noted the Carmen Fair was a prime target for string men since its prize list offered cash for eighty different breeds of poultry! When the prize list was reformed, the "string men" disappeared.²

Parade of Shire draft horses, New Westminster Fair, B.C., 1920s. The man in the lead is Thomas Rollinson, a Shire-horse breeder from Innisfail, Alberta. The Shire was a not-so-common draft-horse breed. One aim of exhibiting draft horses was to promote a particular breed. Canada's fine railway system made such long-distance exhibits possible.

To encourage livestock shows at fairs, governments in Canada gave varying grants to individual agricultural societies. The availability of "free" government money led many fairs to create various methods of profiting from such largesse. Myriad schemes have been concocted over the years to maximize government grants and minimize payouts to exhibitors, thus leaving the society with extra cash for "other things." The Wakaw Fair in Saskatchewan had a good scheme going for much of the 1920s. The provincial government paid half the prize money for this class C fair, so the fair offered $12,000 in prize money, way above its status as a class C society. The top prize was $45, a princely sum in the 1920s. At its peak, the fair averaged 573 livestock entries and 1936 other entries. However, livestock exhibitors were forced to pay an entry fee of one-third of their winnings as well as a percentage deduction of their cash awards. Exhibitors lost as much as half of their winnings to these deductions, but because of the large sums involved, few complained. They still took home excellent prize money even after all the deductions. But the society profited by this scheme. In 1926, the actual payout was

$6,486, while $6,845 was retained by the society. In other words, the society actually made money from the government grants. In 1929, the provincial government stepped in to halt the ruse. In 1907, in Lloydminster, a town on the Alberta-Saskatchewan border, one clever person developed another scheme to milk the grant system for more money. In that year, two fairs were held in town: one on the Alberta side of town, and the other on the Saskatchewan side. Of course, two grants were applied for. Needless to say, the scheme was also halted by government officials.[3]

Altogether, agricultural fairs have played a huge role in livestock improvement and in introducing new breeds. Most Canadian farmers got their first look at new breeds at the local fair. European breeds were being introduced in Eastern Canada by agricultural fairs as early as the 1830s. At the 1909 Calgary Fair, a few Herefords were shown by exhibitors from British Columbia and Manitoba. By 1919, Herefords were dominant at the cattle show. Brandon Fair banned horned cattle by 1928, opening the way for the "polled cattle" only rule. One local dairyman could not capture the coveted "best dairy cow" class at the Armstrong Fair in British Columbia, so he imported a purebred Jersey from Ontario and won the class the next year. His neighbours, impressed by his plan, soon followed suit and imported their own Jerseys from Ontario. Such competition led to an overall increase in milk production in the Armstrong area. Such stories are what livestock competitions and fairs were, and still are, all about.

Judges and Judging at Livestock Shows

No story of livestock shows would be complete without a discussion of judges. Judging has to be one of the most difficult jobs associated with

Six-horse hitch class at Brandon Fair, Manitoba. Brandon Fair was long known as a centre for draft horses. A first-prize ribbon at this fair was really worth its weight in gold among heavy-horse breeders. A fair is really not a fair without its horses.

Parade of Champions CNE, Toronto. In the foreground are the backbone of heavy-horses in Canada: Clydesdales.

the fair industry. The selection of judges was often a problem in the past. In the pioneer era, qualified (and the term is loosely used!) judges were hard to find. Many societies preferred local residents, a trend since discouraged. Often, prominent local officials such as reeves, mayors, members of Parliament, doctors and local merchants were enlisted as judges, simply because they were well known in the area. Competency was not always the measuring stick for selecting a judge. A good example of this was recorded at the London, Ontario, Provincial Exhibition of 1877. One exhibitor of Southdown sheep was preparing his animals when a gentleman walked up and inquired what breed they were. When the exhibitor told him, the gentleman mused aloud, "I believe I am to judge that class today."[3] Needless to say, the exhibitor quickly lost his confidence in the judge's abilities! Another judge sailed through horse classes so fast that the clerk could not record the placings. When asked to settle a dispute over which colt had finished first, the speedy judge was unable to identify any of his placings. This same judge, when he discovered that the breeders' herd cattle class contained only one entry, tried to give the bull first prize, the cow second and the heifer third! Many fairs did not arrange for judges in advance. The fair directors simply hoped that someone would show up on fair day and judge the livestock classes. Many times the directors were forced to desperately scan through the crowd in search of a judge. Obviously, the quality of judges varied dramatically under this system.[4]

Field-crop judging course, Ontario Agricultural College, Guelph, 1908. These students would later become registered departmental judges. The use of trained judges was just one step towards improving agriculture and fairs. The Ontario Agricultural College was a national leader in agricultural education. A degree from the OAC was highly respected all across Canada. It sure looked good on a résumé for jobs or positions related to agriculture.

As is clear, judges were often asked to judge breeds they were unfamiliar with. One such occurrence was recalled by the famous horse judge A.W. Galbraith:

"It has been my privilege to judge no fewer than 17 breeds of horses and ponies at practically all the leading shows on this continent, including the World's Fairs at Chicago and St. Louis and the Panama Pacific at San Francisco, but on one occasion I did feel flummoxed when after judging all recognized breeds of horses at the Minnesota State Fair I was called on to judge the Jacks [a breed of horse] as well. I tried to beg off by declaring my ignorance of the species, but was informed by the management that there was nobody else to do the work. At the noon hour of the last day I wandered into a barn that was full of Jacks, to try and learn something, which I did. The Jacks were all owned by one man, fortunately, so I asked the foreman to show me his best Jack. He pointed to a big plain-looking brute that certainly had no lines of beauty to recommend him. I then asked to see their next best one, and he pointed to a smaller Jack in the adjoining stall. I inquired wherein the first one excelled over the second one, and I remember he said, "Don't you see he is larger?" I replied, "Yes, anybody can see that." Well, he then said, "That's all there is to it, the big one always beats a little one." So I learned a useful lesson at that time and when the class was called those two identical Jacks were the only ones that appeared and I placed them accordingly."[8]

Faced with a tidal wave of criticism from exhibitors, fair directors tried alternative ways of selecting judges. The Kilmarnock Fair in Manitoba decided to let exhibitors vote on judges. Some of the less scrupulous exhibitors hit upon a system to stack the judging lineup. They enlisted friends to make a token entry in a class, thus giving them a vote in judge selection. The schemers then selected the judge who could be counted on to award the prizes to the schemers, regardless of the quality of their animals. Next year another clique of exhibitors voted their lackey as judge. This system did not last long. Firmly convinced of the failure of democracy in the show ring, the directors opted for a three-judge committee. One can only imagine the new problems that were generated by this system![5]

Early Vancouver fairs adopted a novel approach to selecting judges. It was called "trial by jury." Six judges were hired by lottery selection, with four names drawn for each class. Three were to be actual judges, with the fourth acting as a referee. It was not clear if the referee was for the judges or the exhibitors! One spectator sarcastically remarked that the judges sometimes outnumbered the exhibitors.[6]

Sometimes the judges found themselves overwhelmed by the number of entries in the class. One judge at the Regina Agribition found himself buried deep in a class of over fifty cows. Feeling surrounded and needing a better vantage point, the judge declared a time out and returned a few minutes later—on horseback. From this elevated perch, he continued to judge the class. Special problems require special solutions. As he left the show ring, the judge was asked if he "got the class right." He replied, "I don't know if I got it right, but it's official."[7]

Judges often worked under less than ideal conditions. Many early fairs had inadequate stabling or lacked pens for the smaller animals. It was common to judge pigs while they were still loaded in a wagon box covered with canvas to keep out the sun. The judge was forced to either feel the pigs or peek through a knothole or a crack in a board. This was not a good way to evaluate animals. Other fairs saw no problem in allowing spectators in the show area. Spectators and exhibitors often became so thoroughly mixed up in the ring that the judges missed entries. Another problem for judges, and judging was the pressure that some fair directors applied to have them look favourably upon their own entries.

Another problem for judges was the fair prize lists. In the early days of fairs in Canada, classes had only the most general distinctions. For example, the prize list might call for "best bull" or "best stallion." These classes could attract everything from Holstein to Hereford bulls, a totally unfair competition. Some judges simply refused to judge these general classes, knowing the proper way to show animals was to compare apples to apples, or Herefords to Herefords. Another class that caused judges to cringe was "draft stallion of any age or breed." Other problems encountered in the prize list included defining the weight of a heavy draft horse. The standards varied from 1,600 to 1,450 pounds minimum weight. Standardization of ages and classifications was a difficulty, particularly in the livestock sections. Most provincial bodies encouraged standard terms for prize lists, and finally, in 1965, the federal government introduced the "Hays Classifications" for livestock classes. Fairs receiving federal government money were finally forced to standardize prize lists across Canada. It was no play, no pay; a good move for fairs. Although federal funding is now gone, the standardized classifications still exist.

There were so many complaints and problems with livestock judging that eventually government bodies in charge of fairs were forced into action. Some provinces set up judging schools or short courses to train prospective judges in the fine points of their craft. Those passing the courses were designated departmental judges. Their names were placed on a list of approved judges which was sent to all fairs. By licensing judges, it was hoped that the quality of judging would improve and a uniform standard would be utilized. These uniform standards were an immense benefit to both exhibitors and fairs. The livestock exhibitors soon learned what most departmental judges wanted to see in the show ring and they adapted to the new standards. The departmental judge system eventually expanded to apply to livestock, field crops and homecrafts. Ontario had 124 registered judges in 1906 (with 182 fairs), and roughly 50 percent of Ontario fairs used their services. Rates for judges ranged from $5 to $8 a day! So successful was this system that in four years the number of certified departmental judges doubled to 255 in Ontario.[9] Other provinces also adapted the departmental judge system. As in Ontario, the training and licensing of judges was carried out by the Ministry of Agriculture Extension Branch or certified universities. While the judging systems certainly improved, to some exhibitors they were still less than perfect. Exhibitor complaints about judges did not completely disappear. But then, there were always hundreds of "expert" judges at every fair.

To be fair, livestock judging, both past and present, is a difficult and inexact science, as one Prairie newspaper editor explains:

"Animal perfection is such a complex thing, and so impossible of quantitative analysis, that a court of judges can never be organized who will always agree. This judging of two animals is not a matter of one consideration. It is a matter of balancing a hundred considerations. Almost any two horse judges, for instance, will agree as to which horse has the deepest chest, which the strongest back, which the finest feather, which the toughest hoof, which the most springy pastern, which the cleanest bone, which the most typical head and the proudest crest. Point by point they may agree, but when it comes to balancing up the whole matter they disagree. And if they are anything better than slavish scorecard estimators it is quite impossible for any breed association to harmonize them. Their preferences are deep seated and inbred, built upon life experience and individual taste and intuition. And to secure exact harmony in personal tastes whether as to food, as to dress, as to architecture, as to livestock excellence, or as to many other things, is quite out of the question. And judging livestock is and always must be more or less a matter of personal taste." [10]

Tricks of the Trade

Livestock exhibitors always try to make their animals look better in the show ring. To this end, a number of practices have been employed, some perfectly acceptable, some less honourable. The accepted practices are called "fitting" an animal for the show ring, and it is truly an art form. The creatures are washed, their hair trimmed, combed, fluffed and styled. While every livestock exhibitor has fitted their own animals using sundry tricks of the trade, the larger shows or exhibitors use professional fitters. These people are hired to make the entries look more attractive in the ring. A comparison may be drawn with beauticians, hair stylists and makeup artists. Levi Jackson, a professional cattle fitter, described his trade this way:

> "You can dispose of all your clothes, makeup and hair and go naked to show what you really look like. Clothes, hairstyles and the like are things that make a person more or less attractive. Same thing with cattle. You can only improve upon what you have to work with." [11]

Pigs were commonly "powdered" to whiten them for the show ring. This led one exhibitor's wife to grouse that her husband "used to buy more cosmetics for the pigs than I bought for myself." [12]

The dark side of fitting animals involved tricks and chicanery. Some cheaters were very creative in preparing their animals for the show ring. False hair was glued on horse's legs to give them that classic draft-horse look. Another common ploy was to use false manes and tails. One despicable cad even stitched a pouch of ginger under the horse's tail. This induced a rectal burning and itching that naturally would turn the most lack-lustre steed into a snorting

Jersey cattle class, Fredericton Fair. The dairy industry was very important in Eastern Canada. Cheese and butter were among Canada's largest exports until exceeded by Western Canadian wheat. The breeding of dairy cattle is still a major export industry in Canada today.

Grandstand and crowd, Belleville Fair, Ontario.

prancer. One less than ingenious exhibitor of dairy cattle devised a deadly plan; just prior to judging, he pumped milk back into his cows' udders to give them that full appearance judges look for. Tragically, the cows all died after the show from milk poisoning. In more recent years, performance-enhancing drugs have been used on some animals. Horse-pulling associations have now initiated drug tests for such substances. In an era when human athletes use performance-enhancing drugs, it is not unusual for their animal counterparts to be primed in the same manner.

Many livestock classes are grouped by age, but proving the age of the animal has sometimes been a difficult proceess. At the 1902 Western Fair in London, Ontario, one exhibitor lodged a protest claiming that the winning entry in the Fat Steer Class was older than the maximum three years allowed under the rules. An inquiry was called by the board of directors. Three veterinarians were dispatched to examine the four steers entered in the class. The three vets examined the teeth of the steers — the main determinant of age. One favourite method of cheating was to file the teeth. But all three vets ruled the animals to be under the age of three. The prize-winning owner also produced affidavits from witnesses who verified the birth date. The protest was dismissed, and the original placings upheld. Livestock shows and prize placings were taken very seriously indeed! [13]

Horse-jumping show, 1905. The infield ring is crowded with spectators. Some stand, while others cleverly park their buggies close to the ring for the best view. The number and variety of horse-drawn buggies in this photo is amazing. What a traffic jam if they all tried to leave the grounds at the same time!

Horse Shows and Other Four-legged Events

Horse shows at fairs can be roughly divided into heavy- and light-horse classes. The heavy-horse sections may be defined as draft or work horses, and these may be further subdivided into "line classes" and "hitch classes." Line classes are shown on a halter or "lead line." Most of these classes centre on breeding animals, especially mares and foals. The hitch classes comprise horses, usually teams, hitched to a wagon. Heavy horses are shown by breed; light horses are often shown according to the role or purpose they fulfill. The light-horse classes include thoroughbred, carriage, roadster, hackney, saddle and pony classes, to name a few. There have been conflicts over the years about the role of light horses at fairs. They were often branded "pleasure horses" and were ineligible for government grants. One of the stigmas against light horses may have revolved around horse racing. Who has ever seen a race for draft horses or has bet on a line class? Yet light horses are a vital part of almost every fair. For several decades, in the mid-1900s, draft-horse classes virtually disappeared at most fairs. Tractors had replaced the noble horse as the beast of burden. But history moves in circles, and that gallant beast, the draft horse, made a comeback in the last quarter of the twentieth century. Farmers began to show draft horses for a variety of reasons, nostalgia being the main one.

Some of the livestock classes sponsored by fairs throughout the years have been very imaginative. The Saskatoon Fair of 1887 held a class for "best walking team." The prize list did not specify what the team was to

Horse Pull, 1950s.

consist of; so one clever farmer entered his team of oxen. Needless to say, the placid oxen won the class. The horsemen protested, but to no avail. The same fair also had a class for "best trotting ox, hitched to a buck-board." Now, the term "trotting ox" is an oxymoron. "Lumbering ox" might be more apt. The class was soon discontinued. Moose Jaw Fair once held an ox-team race. The winning time was not reported, although it should be pointed out that it was only a one-day fair.

Local businesspeople were always canvassed for donations for horse show prizes. While cash was the preferred medium, special donations tied to the donator's business expertise were not uncommon. At one early Edmonton Fair, the prize list contained the following special prizes:

Best team of horses: $25 in legal fees, to be used for divorce proceedings only.

Best display of grain: $25 in doctors' fees for amputations only.

Best loaf of bread by a bachelor: free marriage ceremony: five-year limit, interdenominational.[14]

In the age before tractors, draft horses and oxen were the beasts of burden on farms. They did most of the plowing and hauling and heavy work. A good team of either oxen or draft horses was often the farmer's prized possession. It was only natural that farmers would show off the drawing prowess of their prize teams at the local fair. Thus the "pulling contests" or "drawing matches" were born. In their most common form, these drawing matches involved a team pulling a weighted sled a specified distance in competition with other teams. The sled was called a "stone boat" after the simple devices many farmers used to haul rocks and stones from their fields. In later years, the stone boat was loaded with concrete blocks instead of stones. The specially formed concrete blocks could be handled more easily and were more evenly weighted. In the pioneer era, oxen were common, and many fairs had classes for both oxen and horses. In the twentieth century, oxen became scarce in most regions of Canada, and the draws evolved into horse pulls. However, in Eastern Canada and the United States, oxen are still common, and ox pulls are still held at many fairs.

Oxen and horses are trained differently for pulls. In a horse pull, the teamster hitches the high-spirited animals from behind the team and guides the pull with reins, never letting the team see him. In an ox pull, the teamster leads the team by walking in front. He or she guides or encourages the team with a whip, although most teamsters never really whip the animals. Many trained ox teams will not pull unless they can see their teamster. Horses will jerk or jump when they start their pull. Oxen tend to "lean into" the pull. Both types of beasts exude a certain class or charm when they pull. There is a quiet grace to a horse or ox pull that is missing in a tractor or truck pull. These animals, although specially trained to be farm pulling teams, are often employed in other jobs such as logging, doing farm work and pulling wagons to haul passengers. It is mandatory that pulling oxen have horns. The ends of the horns are usually tipped with round metal protectors as a safety feature. Since oxen are overgrown or aged steers, the most common cattle breeds appear in the ox pull. And ox pulls, like horse pulls, are often divided into heavy and light classes based on size or weight of animal.[15]

Ox pull, Bridgewater Fair, Nova Scotia. In the Maritimes, ox-pulling contests are still common, long after such classes have disappeared in the rest of Canada.

Heavy horses, on the other hand, are usually represented in Canada by three breeds: Clydesdale, Percheron and Belgian. Other breeds included under the heavy-horse designation are Shire, Halflinger, and Commercial or Draft Ponies. Hitch classes most often do not have a designated breed and are open to all. The light-horse division is much more complicated. The following classes are offered at a typical mid-sized fair:

Hackney Horse	Western Saddle Horse
Roadster	English Saddle
Hackney Pony	Purebred Arabians
Local Pleasure Horse	Morgan Horses
Shetland Pony	Quarter Horses
Welsh Pony	Hunter/Jumper show

These are just some of the classes included in the light-horse division. And these light and heavy horses were not the only equine creatures at fairs. Racehorses were common, and any fair that sponsored a rodeo also included dozens of horses in various rodeo classes. See the "Entertainment" chapter for these.

Top to bottom: Roadster/carriage class at the CNE. In the era before motor cars, many urban residents kept a carriage horse on the premises. A well-turned-out carriage was a source of pride; and what better place to show off than at the local fair. A first prize in this class really meant something!

Lady Riders, saddle-horse class, 1923. It was still unladylike to ride horses straddle. Formal dress and side-saddle was how a "proper lady" rode. Gentleman grooms were also present to render assistance and add a touch of class to the show. White dresses seemed to be the uniform of the day.

Ayrshire-cow class, Sherbrooke Fair, Quebec. Before 1920, Ayrshires were the most popular dairy breed in Canada. A sign on one cattle barn announces Jersey cows, another popular dairy breed at the time. Gradually, Holsteins took over the popularity contest in the dairy industry. Notice the handlers, immaculately attired in white.

Cattle

Nineteenth-century fairs featured only a few breeds of cattle. Shorthorns dominated the beef division. By 1900, Herefords and Angus were becoming common. By the 1960s, a new wave of European breeds began to flood the livestock market. Fairs responded to these changes by adding classes for Charolais, Blonde d'Aquitaine, Limousin, Simmental, Salers, Belgian Blue, Pinzgauers and even the woolly Highland cattle. It is not unusual for one fair to offer prizes for up to eight beef breeds. Add to the breed classes such departments as market cattle, market steers, 4-H shows, etc., and the fairs can be awash in beef cattle. The dairy shows at early fairs were dominated by Ayrshires. Jersey, Guernsey and Holsteins were introduced later. These dairy breeds still dominate today, although Holsteins are by far the most common breeds.

Ayrshire dairy class, Sherbrooke Fair, Quebec, 1912. Notice the long horns on the animals.

The "Smalls"

Often lost in the shuffle of livestock at the fair are the smaller or lesser-known or even "less romantic" farm animals. Sheep, pigs, chickens and

4-H Dairy Cattle Show, Port Perry Fair, Ontario, 1950s.

other poultry, goats, rabbits and many other breeds of domestic farm animals all have their place at the fair. The same motives that led farmers and livestock owners to bring their big animals to the fairs also inspired the same farmers to haul a myriad of their smaller animals, along with them. Elkanah Watson, the originator of the Berkshire system of fairs, started these fairs to encourage farmers to produce more and better-quality wool for his woollen mills. Sheep were his main focus.

In the past, most farm operations were general or "mixed," where farmers grew many different crops and raised many different animals on their acreage. They were likely to have horses, cattle (both beef and dairy), pigs, sheep and poultry in large numbers. The children's song, "Old Macdonald's Farm," with its "chick chick here" and "moo moo there" was certainly a "mixed farm," and the keeper of such a farm could, if he desired, bring quite a selection of animals to show at the fair.

Goat Class

Sheep Class

In the last fifty years, mixed farming has gradually been replaced by specialized farming. Modern farms are more likely to raise only one or two breeds of livestock or field crops. Once, every farmer in a township raised a few pigs or kept a dairy cow; today, the nearest pig or Holstein may only be located many miles away. Pigs, especially, are now largely produced in large factory farms: very few farmers keep breeding pigs on their premises. Thus, the availability of many livestock breeds has severely dwindled. As a result, some fairs have trouble attracting pigs or sheep or other livestock breeds. Swine classes have disappeared at many fairs, and are often replaced by simple displays. It is still widely felt that pigs must be presented to the public fair, and displays are one practical way of doing this.

Fairs have responded to these community changes in many ways. Some fairs simply cancelled certain classes. Others tried to attract exhibitors from far away by offering more prize money, travel bonuses or by holding special "point-shows" to attract exhibitors. (Livestock breed associations passed out "points" based on prizes won at sanctioned fairs. At the end of the fair season, these points were tallied and a grand champion declared based on the results.) Still other fairs simply paid farmers to put on non-competitive displays of various animal breeds. Unable to attract enough animals to put on a proper competitive show, the directors of a society kept a "proper tail on their kite" with these educational displays instead. Showing a pen of pigs or sheep, one dairy cow from each breed, or an assortment of poultry were all good ways of presenting the farm story to audiences. In many areas of Canada, urban-based patrons far outnumber rural fair-goers. Show ring competitions usually mean little or nothing to the urbanite. An educational display of farm animals is often better understood and more digestible to those from a non-farm background.

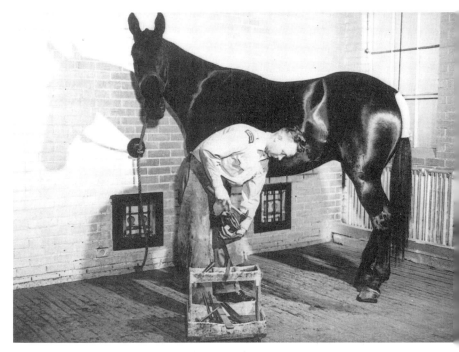

Horseshoeing display, Brandon Fair, Manitoba. Blacksmiths, or farriers, were once common, but in the last half-century, not many Canadians have ever seen a horse shod. Thus, many fairs began to put on demonstrations of horseshoeing, cow milking, sheep shearing and other disappearing farm skills. Fairs have dipped into their Canadian history to educate Canadians on what rural life was once like.

These displays usually contain snippets of information on the breeds, baby animals and mothers are featured, giving the display a cute, folksy atmosphere that young families can relate to. Large numbers of one breed are eschewed in favour of one or two from other breeds. Some of the larger exhibitions produce really elaborate displays. In many cases, the viewer is allowed to pet or touch the animals. Sometimes pregnant animals are imported in the hope they will give birth *at the fair*— an extremely popular occurrence and on other aspects, such as where milk comes from. Thus, the competitive show ring has given way to the educational display at many fairs.

The Petting Zoo

Another method of introducing kids and animals at the fair is through the petting zoo. A hundred years ago, the petting zoo was unknown. Many families had farm animals at home or knew families that did, and children were totally familiar with animals of all descriptions. Most had their own petting zoo right in their backyards. But in today's world, it's tough to keep a pet pig on the tenth floor of an apartment building. Likewise, one no longer needs a carriage horse to go for a Sunday drive. The number of animals, especially farm animals, readily accessible to urban families, has shrunk dramatically.

Petting zoos have moved in to fill this void, and here, children can see and learn about farm animals of many types. The emphasis on baby animals and petting or touching them makes petting zoos extremely popular with young children, a clearly favoured target audience at fairs. The petting zoo also meshes well with the livestock competitions that adults love. After all, a fair without animals is like a kite without a tail.

In recent years many fairs have moved into classes for "non-traditional animals." Some of these classes are serious breeding classes, while others are "fun shows." Dog, mutt or pet shows have become popular recently. While some larger fairs sponsor serious purebred shows, most pet shows fall into the fun category. Proud pet owners bring their animals to show off and have a bit of fun. The pet show allows the non-farm resident to participate in animal shows at the fair. It's no longer necessary to have a prize calf or saddle horse; the common household dog or cat or guinea pig will do. And if the animal can do some sort of trick, albeit poorly, so much the better. After all, fairs are supposed to be fun!

Petting Zoo

Some of these nontraditional fair animals actually have Canadian farm roots, such as rabbits and dairy goats. Others, however, are a little out there. Miniature animals, for instance, are often put on display. And even such animals as llamas and alpacas have crept into many fair line-ups. The following prize list for llamas and alpacas from the Millarville Fair in Alberta mixes both excellence in breeding with humour and fun—a combo that is common to the Canadian fair.

Section 53: *Llamas and Alpacas entry rules*

1. Must enter with the intent of having fun!
2. No flirting with the judge will be allowed.
3. No touching, kissing or hugging except to llamas.
4. Only gentle pinches to the Ring Master will be tolerated.
5. All bribes must be done discreetly through the club treasurer.
6. All mumbling, grumbling and complaining must be written on the back of a postage stamp (for easy return) and submitted to the grumbling committee for evaluation. Awards may be given for the best ones received.
7. Handlers better looking than their llamas not allowed.
8. No foul language, insults, stealing, etc. will be tolerated unless agreed to by the other party or parties involved.
9. When you see someone without a smile you must give them one of yours.
10. All handlers in the show ring without a llama will be asked to leave the ring.
11. No boy llamas prettier than girl llamas allowed, unless okayed by the judge.
12. No prizes will be given to people caught disguised as a llama.
13. No spitting in the show ring except by llamas.[16]

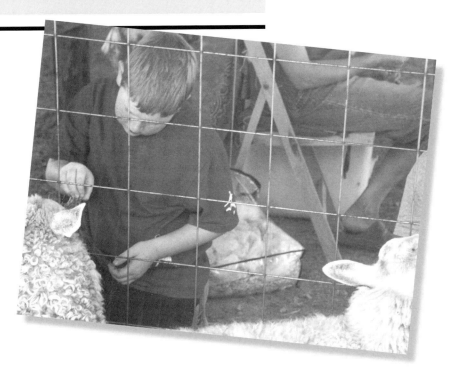

Petting Zoo. Look at the intent expression on the boy's face. Is this concentration or what?

Livestock is the most common denominator of agricultural fairs, the real centrepiece. It was in the past and so it is today. The breeds and classes may have changed over the years, but the presence of farm animals at a fair is still a must. After all, a kite without a tail is not really a kite.

Young girls' horse-hitching contest, Peel County Ontario School Fair, 1917. The crowd seems engrossed as a young girl hitches a horse to a buggy; a useful skill in the era when everyone in rural Canada travelled by horse and buggy. The girl is a little overdressed for such a chore, but judging by the crowd, everyone dressed up for this school fair.

THE LIVESTOCK RING AND THE BEEF CATTLE INDUSTRY

Livestock judges at fairs have always had a big influence on the livestock-breeding industry, and nowhere more than on beef cattle. In the pre-1920 era, the beef industry was dominated by Shorthorns, with Hereford and Angus distant competitors. The Shorthorns were big-framed animals that flourished in Canadian climates, did well on rough forage and produced large, rough carcasses at the packing plant. The establishment of the Royal Winter Fair in Toronto in 1922 gave the livestock industry its grand champion show so long desired by cattle breeders. Early judges at the Royal were from the meat-packing industry. In the open-market classes, these judges, to the dismay of Shorthorn breeders, began to favour Angus steers. They maintained that the blockier, smaller, more compact Angus cattle could be cut into steaks and roasts more efficiently. The subsequent show-ring results led to a rush of breeders abandoning Shorthorns in favour of Angus and later Herefords.

The move toward smaller animals quickened during the Great Depression, when the demand for beef dropped and the U.S. beef export market collapsed. As show-ring judges continued to reward smaller, more compact animals, beef breeders scrambled to produce even smaller, shorter-legged, compact animals. Beef cattle were actually downsized in the 1930s. After World War II, larger cuts and carcasses came back into fashion. The judges led the way by rewarding the bigger, beefier animals. Beef cattle were "bred up" in size. In the 1980s, the demand reversed and leaner carcasses came back into fashion. This dizzying pattern of change and demand was acted out in show rings all across Canada.

Hereford bull. Notice the short legs and the horns, both traits discouraged today.

In the rush to find bigger, more quickly maturing carcass beef cattle, Canadians searched abroad. By the 1960s, new breeds of beef cattle began to appear at Canadian fairs. The Charolais from France started the exotic beef breed influx. They were followed by Simmental, Limousin, Maine-Anjou, Blonde d'Aquitaine and, most recently, Belgian Blue, Salers and Galloway. The advantages and traits of these breeds is a subject to itself. But it is safe to say that most Canadian farmers were introduced to these exotic breeds at a fair. As one cattle breeder explained:

"The fairs ... got breeders looking at each other and there were the exotics, way the hell and gone up there in carcass sides.They presently realized that to compete they had to get to their calves up there too. The Angus had to get up there. My God, it has changed. It changed because people got to [fairs] and they talked."[1]

M.M. BOYD AND POLLED HEREFORDS

One of the greatest of the genetic advances in the livestock industry in the twentieth century has been the increasing dominance of polled cattle. In 1900, horned beasts were the norm. Polled or hornless cattle were considered genetic "freaks" or "sports."

Cattle with horns had several disadvantages. The horns were dangerous to other animals when used as weapons in fights. They were also dangerous to the farmer. Sometimes the horns grew into the side of the animal's head and could kill it. Most farmers cut off the horns, but this was a painful procedure that often harmed the animal's health. Then in 1859, Charles Darwin published his landmark book, On the Origin of Species. Progressive farmers read this revolutionary work and began to apply Darwin's principles to livestock breeding. They realized that livestock could be bred to produce certain desirable characteristics, including size, temperament, body structure and lack of horns.

In the 1890s, a prominent Ontario lumberman and erstwhile farmer named Mossom Martin Boyd became interested in animal breeding. Boyd was a lumberman operating from Bobcaygeon, Ontario, but like many gentleman merchants of his era, he had a strong interest in agriculture. He began to experiment with crossbreeding and animal genetics. He dabbled in Angus, Shorthorns, Durham and Galloway breeds. Boyd also carried out a lengthy attempt at crossing buffalo and domestic cattle. It was hoped that these hybrids would have the native buffalo's hardiness for the Canadian climate and the fine meat of European cattle. It all started by chance.

In 1886, an American named D.C. Winston shot a buffalo mother on a hunting expedition to Dakota just as the last of the great herds were driven to extinction. Standing beside its mother's body was a young buffalo calf. Maybe a tinge of guilt afflicted the hunter, for he named this buffalo calf Napoleon and took him home to California as a pet. Here, Napoleon became a fixture at a large hotel. Mossom Boyd saw "Bonie," as he was nicknamed, on a visit in 1893. Boyd then purchased the buffalo and began crossbreeding experiments with domestic cows. The resultant "cattalo," as they were called, never displaced the standard cow as a beef animal. Mossom Boyd, however, tried, and used to take his cattalo to the local fairs to show off his genetic experiments. For years, Napoleon the Buffalo led the Bobcaygeon Fair parade. His mounted head is still displayed in the Bobcaygeon library. When Boyd died in 1915, the remaining cattalo were sent to a Saskatchewan park.

But the most important offshoot of Mossom Boyd's crossbreeding programs lay with polled Herefords. He communicated with like-minded breeders in the U.S. who were also searching for freaks or sports. In 1903, Boyd purchased two purebred polled Hereford bulls and began his breeding program. Boyd transferred some of his purebred herd to his new Saskatchewan farm. There in 1912 was born, arguably, the most famous polled Hereford bull in Canada: Bullion 4th. In 1913, Bullion 4th won the Dominion Grand Championship for Hereford bulls. To the Boyd family, winning championship ribbons at major fairs was the pinnacle of success. Their herds toured the fair circuit. Bobcaygeon had no railroad, so the Boyd Herefords were taken by boat to Lindsay and then to a train to major fairs such as those in Ottawa, London and Toronto, and also, many shows in the U.S. While auctions and breeder magazines were important places to advertise the polled Hereford breed, it was the fairs that provided the greatest exposure. Farmers and cattlemen from all over Canada could view the breed in the show ring and at the cattle barns. It was vital to the Boyd family to show their stock at as many fairs as possible. The fair was both marketplace and classroom.

Through the efforts of enlightened breeders such as Mossom Boyd, polled cattle went from freaks to the norm in the beef cattle industry. It became virtually unacceptable to have horned cattle at a fair. Thanks to the efforts of progressive breeders and agricultural societies and fairs, Canada became a world leader in animal breeding. Canadian livestock has become among the best in the world. Breeders from all around the world attend Canadian fairs and livestock sales to buy first-class breeding stock.

Cattalo. The breed never caught on with farmers.

Parades

EVERYBODY LOVES A PARADE. A century ago, parades were just as popular as today. May Day, Labour Day, Canada/Dominion Day, Christmas, Thanksgiving, Armistice Day, wartime drives, even special one-time events such as sports championships, historical anniversaries and visits from dignitaries were all good excuses to have a parade. Of course, the annual fair also fit the parade-excuse list very neatly. Fair parades had unique features. They were likely inspired by circus parades in which all the animals and actors marched from the railway station to the circus site. This was done to advertise or promote the circus. While fair parades were organized to entertain the public, they were also designed to promote the fair and encourage the viewer to want to attend the event. Many fairs schedule their parades to kick off their event.

No fair parade is complete without animals, especially the noble horse. Before cars, each community or town would contain hundreds, even thousands, of horses. They were kept for a variety of uses, for work, transportation, even pleasure. At fairs, horses were necessary to draw the floats, wagons, etc. But in the age of the motor vehicle, horses became either an exhibit or a piece of nostalgia. Even as early as 1912, the directors of the Edmonton Fair, for instance, made the theme of their parade "the forgotten urban horse." Four hundred city workhorses joined the parade for the salute to the workhorses that made the city of Edmonton run. The winner of this class was named "horse of the streets." A special class for horses over eighteen years old attracted twelve entries. The Jock and Pete were the winning team of urban draft horses; they were twenty and twenty-one years old respectively. The theme concept was a way to let people know how important the average or general-purpose draft horse was to early-twentieth-century urban society. They pulled everything from delivery vans to fire wagons.

Another unique feature of the fair parade was its commercial advertising. Businesspeople and manufacturers outfitted large and elaborate floats to promote themselves and their wares. The ready-made crowds made ideal audiences for their marketing. The variation of goods and services promoted staggers the imagination. The creativity … well, creativity is in the eye of the beholder. One picture is worth a thousand words. So here comes the parade!

Government of Canada entry in a fair parade in Exeter, England, 1907. Part of the government's policy was to promote immigration to Canada as a whole and to the Prairies in particular. The local immigration agent conceived the idea of entering a small float in the local fair parade. The decorations are a curious blend of Canadiana and Oriental mystique. It was hoped such advertising would inspire prospective immigrants to think about moving to Canada, instead of, let's say, the U.S. Such advertising campaigns had astounding success, for over one million Britons immigrated to Canada between 1890 and 1915. Oh yes, our gal won first prize, even with the Australian looking hat and Chinese lanterns!

Warriors' Day Parade, CNE, Toronto. Every year the CNE opened with a huge Warriors' Day Parade led by massed pipe bands. Military men, past and present, paraded in the event. In the background is the famous Princes' Gates, named in honour of two princes of England: Edward and George, who officially dedicated the entrance in 1920. The Princes' Gates is a landmark at the CNE to this day.

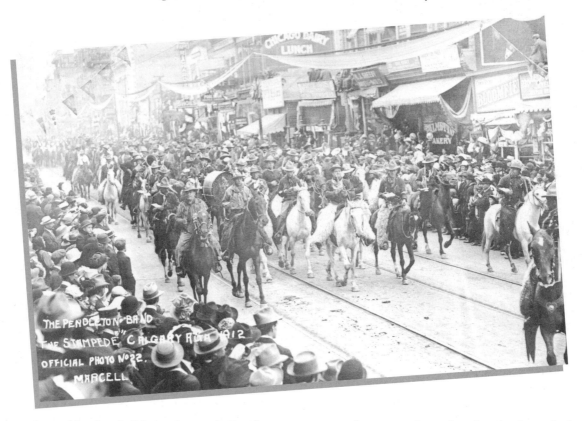

A mounted marching band, Calgary Stampede Parade, 1912. The Calgary Stampede was noted for doing things a little differently. It was not often that a parade contained a mounted marching band. I hope the horses didn't mind the sound of tubas or bass drums.

Bobcaygeon Fair parade, Ontario, 1908. In the era before plastic decorations and fluorescent colours, nature's finest fall foliage was the best decoration to be found.

Icelandic Canadian float, Calgary Fair parade, 1908. Participation by cultural groups was part of many fairs both past and present. The banner reminds Canadians that Icelanders (Vikings) were the first Europeans to set foot in North America circa AD 1000. It would be another 872 years before another Icelandic settlement appeared in Canada, at Gimli, Manitoba, after an unsuccessful year-long stint at Kinmount, Ontario. This float won first prize at the Dominion Exhibition Parade. I hope they didn't try to tow (or row) this float in the parade, as it is set on wooden sawhorses in this posed photo.

A forty-horse team in the Calgary Stampede Parade, 1925. This long team was assembled as a special attraction. It hauled ten wagons loaded with barrels of grain or spirits. It also needed a few outriders to clear the path through the crowd. How would you like to steer this hitch around a street corner? The huge crowds lining the route made the outriders necessary for crowd control.

A float advertising "Robin Hood Flour," Vulcan, Fair Parade, Alberta. A little imagination and a few flour bags were used to promote this popular product.

Ship float, Calgary Exhibition, 1908. Even in landlocked Alberta, Britannia ruled the waves. Businesses in the background include the Imperial Bank of Canada and the Hudson's Bay Company. Patriotism and fairs were like bread and butter.

North West Mounted Police float, 1905. In honour of Alberta's inauguration as a province, the North West Mounted Police built this float. The forerunner of the RCMP was a symbol of Canadian culture, especially in its birthplace, the Northwest Territories. The float is packed with symbols of Canada and the British Empire. No doubt the float received a loud round of applause from the appreciative crowd.

Barnum Bailey Wild Animal Show, historical pageant parade, Dominion Exhibition, Calgary, 1908. It looks as if a white Siberian tiger is on display in the big parade, and likely at the fair as well. For a special occasion fair such as the Dominion Exhibition, no show was too expensive. So the famous Barnum Bailey Circus was brought in.

Above: **Cardston Fair, Alberta, queen float, 1903.** The theme was "Canada and the Provinces." It appears the contestants were attired in dark-coloured wigs and held shields representing the coats of arms of the various provinces. In 1903, Alberta was not yet a province. Maybe Alberta and Saskatchewan are represented by the two young girls in the front, provinces-in-waiting.

Below: **Canadian National Railway float, Calgary Fair, 1914.** So successful was the Canadian Pacific first transcontinental railway in opening up Western Canada, that a whole series of new railway lines soon criss-crossed the country. Canadian National was the second transcontinental railway line.

John Bull float, Dominion Exhibition parade, Calgary, 1908. John Bull was a cartoon that represented England. The float's slogan reads "Britannia Rules the Waves." The military might of the British Empire is symbolized by the military representatives riding on the float.

Simpsons Department Store float, CNE Parade, 1928. The Simpsons chain began to frequent the fair circuit after taking a page from a longtime rival, Eaton's.

WINTER FAIRS

In the beginning, all Canadian fairs were held in the fall or spring. Fall was the favoured season, for it was the time of year when crops were harvested and farmers could celebrate a successful year. It was also no coincidence that the fair season officially ended at Thanksgiving. Anyone in Canada can realize how impractical it is to hold an outside fair after mid-October. Yet many farmers, especially livestock breeders, felt the need for a year-end show after the fair season was completed. There was also the desire for a grand-champion show where all the breeders could come together and compete. Several centres across Canada also became home to fat-stock shows, where livestock was bought and sold at the end of the fall season. For many years, the provincial exhibitions filled this role. But the demise of the provincial/dominion fair circuits left a void for many serious exhibitors, especially in the livestock industry.

It was really the spread of the railway network that made the regional winter fair possible. As many Canadian exhibitors began to travel to the Chicago Fat Stock Show in the late 1800s, a demand arose for similar shows in Canada. Thus began the winter fairs.

The winter fair was designed to be a purely agricultural fair or a farmers fair. There were no midways, no sideshows (at least outside the show ring and the stables), no fakirs or spielers and no hootchy-kootchy shows. In this sense, winter fairs were a throwback to the original fairs of the early 1800s. They were designed for agricultural sales and education only. Winter fairs were held after the fair season was over and before Christmas. They were held entirely inside, thanks to the severity of the Canadian weather at this time of year. It was the final show for many livestock breeders who had toured the local fair circuit all summer and fall.

There were never very many winter fairs across Canada, because, at this time of year, people preferred large, multi-regional competition. Also, during hockey and curling season, there were not that many venues to choose from. The Atlantic Winter Fair in Saint John, the Royal Winter Fair in Toronto, Brandon Winter Fair in Manitoba, and the Pacific Winter Fair in Vancouver were the bigger regional shows. Agribition in Regina was a latecomer to the circuit, holding its first show in 1971. Many Western Canadian livestock breeders, who normally attended the Royal Winter Fair in Toronto, started Agribition to provide a show closer to home and save on travel.

The Royal Winter Fair is the crowning jewel of winter fairs in Canada. It was started in 1922 in the Canadian National Exhibition grounds as a grand finale for the fairs circuit and a year-end livestock-sales opportunity. But it quickly became much more, as horse shows and international equestrian competitions were added. A red ribbon at the Royal represented the pinnacle of success for any livestock breeder. Showmen travelled from all over North America for a crack at a Royal championship. Buyers came from around the world to shop for livestock, and the Royal became the greatest gathering of livestock buyers, sellers and exhibitors in Canada. It also became the biggest agriculture classroom for schoolchildren in Ontario. The Royal is truly a wonderful throwback to the fairs of a bygone era.

SPRING FAIRS

Spring fairs were usually held in April or early May. They were purely business affairs, unlike the fall fair, and were never as well attended as fall shows. The spring fair was designed to promote and market livestock and seed grain. They harkened back to the old market fairs of England, when a draft-horse class is in the ring and snow is on the ground.

One such spring horse fair was described as follows:

"The spring fair was held in Mount Forest the third Wednesday of May. The May fair was where the area farmers were selling and buying horses. The buyers from the city arrived in town by train the day before. The main street was the scene of the action with hundreds of horses on display. The buyers wore their traditional fur coats, carrying a cane, a pocketful of cash on one side and a bottle of whisky on the other.

"I recall the haggling between buyer and seller. Some of the language a young boy wasn't supposed to hear! The owner would lead a horse on the run, to prove there was no shortness of breath, and the horse did not have a common complaint known as the 'heaves.' Prospective buyers checked the teeth for age; the eyes for blindness.

"Then came the offer from the buyer. This usually brought on a heated argument, with the owner walking away. In a matter of minutes the 'city slicker' would up the offer, finally closing the deal with a handshake. Payment was in full in cash. No other paper changed hands. This was known as 'a gentleman's deal.' The buyer led the horse away.

"Main Street was literally packed with horses, buyers and sellers, all haggling at the same time. The hotels were busy places, the stables providing feed and accommodation for the horses; the hotel providing full course meals at 50¢.

"The town on these occasions was a popular place for the roaming gypsy bands, traveling through the country in horse-drawn caravans. Horse trading was their means of survival. And the travelling medicine man was a common sight on fair day, selling medicines guaranteed to cure any and all diseases. The price 50¢ for a large bottle of magic cure, said to be pure alcohol flavoured with lemon. I recall many staunch prohibition ladies later praising the quick results of the wonderful medicine, not to mention the hangover."[1]

Spring fairs gradually disappeared. They were money losers, and most of the functions were usurped by other branches of the agricultural industry.

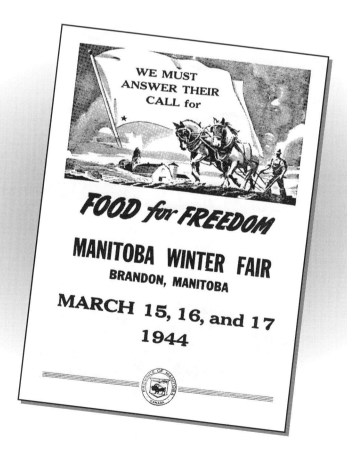

The Thrill of It All: Children and Fairs

A primary goal of every agricultural society is to involve youth in its activities. Almost every fair director or volunteer remembers childhood experiences at the fair. Many readers can also pause and cast their minds back to memories of childhood days spent at a fair. These happy experiences often inspired active participation in local fairs later in life. Or, as one old director used to say about recruiting fair volunteers, "Get 'em hooked while they're young!"

There are many ways to encourage youth participation at fairs. Special junior exhibit classes are one obvious way. Sponsoring 4-H clubs or junior farmers clubs can also develop children's love of fairs. Many midways have a special Kiddyland and special pricing on children's days. Many fairs sponsor children's entertainers and children's activities. Whatever methods are employed, children and fairs belong together.

The School Fair

In the early 1900s, rural children had the opportunity to live the fair experience through school fairs. The school fair was designed for school-age children to participate in their own mini-fair. These fairs

School parade, Bruce County, Ontario, 1919. The various township schools have been assembled for the county school fair. The banners proclaim the townships and school sections represented. The schoolchildren are costumed in theme, ranging from Scottish traditional to English knights to formal white dresses. The students always marched in a parade to start the school fair.

The schoolchildren of SS #6 Sullivan Township school, near Owen Sound, Ontario, at Bayfield Fair. Their theme: "The Farmer Feeds the World: Farming is the Greatest Industry." This would have drawn a smile at any rural fair anywhere in Canada.

were organized by school sections/townships/counties. It was hoped the children would become interested in learning more about rural life, and be introduced to competition and trained in farm subjects. Ontario led the way in school fairs. The first one in Ontario was held in Waterloo County in 1909. Three rural schools and fifty-eight children participated. By 1919, there were 357 school fairs across Ontario involving 3,278 schools and 79,000 children! Needless to say, they were a runaway success. In 1912, school fairs became part of government policy and received grants from the provincial department of agriculture. The school fair introduced many kids to the thrill of competition and the joy of the local agricultural fair. Many exhibitors and fair-goers got their first taste of fairs at the local school fair.

The following account shows the trials and tribulations of staging a school fair. This account comes from Downey Township in Perth County, Ontario, which held its first school fair in 1915.

"Each of the township's nine school sections was represented by a teacher and a director. The directors put up $12 with another $7.25 collected in donations, plus $10 from the township council and came up with a budget of $29.25 with which to stage their first fair. They also drafted a prize list with 25 classes. The site chosen was a wooded grove: the date was Saturday, September 25, 1915.

"Fair day was not without its moments. Two teachers were at the grove by 9 A.M. to receive entries and to display them in the tent that was to be erected by 9:30. But the tent did not arrive until noon, and then the wagon on which it was being delivered from the Stratford Armouries got bogged down in the swampy lane leading to the grove. In fact, almost all the horses and buggies on that day suffered a similar fate.

"President Mountain supplied the lumber, as he was charged to do, but there were no hammer or nails to help turn it into benches and tables. Nor was there anything on which to serve the ice-cream, or wash dishes. The teachers hurried to the store where they bought some nails and towels and borrowed a hammer and a dish pan. A student was dispatched to Stratford with his cart and pony to rent some dishes and spoons. The round trip was 14 miles and his mission was successful. A teacher nailed planks across tree stumps to create the booth.

"Most of the students, their parents and an assortment of fruit, baking, sewing, flowers, calves and chickens showed up before the tent was in place, but there was a mood of understanding and excitement. There were prizes for the best exhibits and for the sports contests. And there was a baby show. At day's end the teachers doled out prizes while the tent came down around them. It had to be back in Stratford by 6:00 P.M."[1]

School parade, Brant County, Ontario, 1925. The children from a one-room school pose before the big parade. The girls are dressed as nurses, the boys as sailors, a curious combination.

By 1922, all eleven townships in Perth County were holding school fairs. The county fair at Stratford realized their value to youth and offered a division for township-school fair winners. When the Perth County school fairs disappeared, the Stratford Fair filled the void.

Some of the classes offered at school fairs were quite predictable shows of crafts, produce and animals. Others were more unique. The Clark Township School Fair in Orono, Ontario, sponsored a class for a collection of groundhog tails mounted on cardboard with a note under each showing the date and a description "how captured." A total of 3,457 tails were shown in 1933. The peak year 1936 saw 7,009 tails on display. The top three students bagged totals of 92, 72 and 63 tails. These were dangerous times for groundhogs![2]

Sunderland, Ontario, resident Jean Baker recalls typical childhood experiences at the local school fair.

"I showed at each school fair from 1928 to 1935 as my parents were very much interested in fairs. We lived a half mile west of town. I enjoyed preparing articles to show. I have a drawing of a banana which I did in first-class and won first prize: my teacher was Evelyn Kaye. The crayons I used were a previous type to the old wax crayons. I can remember drawing and writing were very much stressed in school work.

The school fair was held in Sunderland's old wooden rink. The different school sections in Brock township's Ontario County came together in September for the rural school fair. Sunderland was SS #13. The old rink had a dirt floor with wooden tables set up in the middle, and a bench went all around the edge of the rink to display the different classes of exhibits. Paper plates were used for small vegetables and baking. For a period in winter months this building was used for skating and hockey. In the spring of the year the school gave out one package of flower seeds and one package of vegetable seeds to each pupil interested in the fair in the fall. I think the Department of Agriculture was behind it. I thought the flowers called Salipiglosis were out of this world for colour. Carrots Chantenay long and Detroit dark red beets were the usual vegetables.

"The last year I was eligible to show was in 1935. I put a big effort forth to show as much as I could to win the most points. I won the most points and received a silver-handled cake plate, grape design, donated by the T. Eaton Company. My name was engraved on the centre of the plate for winning the most points at Sunderland Rural School Fair 1935. I won a number of first prizes, one for playing a Hawaiian guitar. I could only play one piece, 'Star of the East,' as I had just started taking guitar lessons. I couldn't even tune it so I had to make the trip down to the minister to get him to tune it that morning of the fair. I made bran muffins that morning too. Dad had

To the Public and Separate School Teachers of Glengarry County (Ontario)

The Directors of the Williamstown Fall Fair are inviting the Elementary Schools of the County to participate in Demonstrations at the 1943 Fair in Williamstown on September 15.

In extending this invitation the Committee is acting in accord with wishes of the Departments of Agriculture and Education. These Departments are anxious that pupils may participate in local fall fair activity as much as possible in order that the educational loss through the temporary discontinuance of School Fairs may in part be made up.

It is the hope of the Williamstown Fair Board that a large number of schools will participate.

Murdie McLennan
Secretary.

School Fair Rules – 1923

The object of the school fair is to create an interest in Agriculture among the pupils of the rural schools, to awaken them to the pleasure and possibility of rural life and to introduce the best seed available.

The pupils shall be given the seed of the crop chosen in the list which they will take home and sew according to directions. They shall care for and harvest the crop and show it at the Fair, according to the directions given for each class. The pupils shall have the same seed and therefore, the same start. The eggs for hatching are from the bred-to-lay Plymouth Rocks originated at the Ontario Agricultural College. These eggs will cost each pupil 80 cents per dozen, and the money must accompany the order. Stamps will positively not be accepted.

Rules
- Teachers are asked to kindly see that the work is that of the exhibitor.
- Whether the results are good or not, each pupil taking seeds must exhibit.
- Keep a careful record of dates and seeding, yield and other items of interest.
- Calves and colts must be made to lead and must show on the halter.

decided to clean the stove pipes on the wood stove and in the hurry the pipes were harder to get together. Time was a problem as entries had to be in by a certain time to be judged. I tried my hand at making a wooden chicken feeder with a spin centre. I thought we had some pretty good mangolds, sugar beets, so I got them from the field, either three or five, washed them, trimmed them nicely and laid them on the back kitchen step to dry. Well, didn't a hen fly over the fence and give them a peck or two, but as time was running out I showed them anyway and got first. I showed brown hen's eggs, also a Barred Rock hen and cockerel. For the sewing part, I hand-hemmed a linen tea towel and still have it. I can remember going to Port Perry Town Hall to play the guitar. Winners of local fairs played off there. I came home with a pen and pencil set.

"My brother Robert showed a calf which he was to lead around on a rope. The calf decided not to move so his friend Don Christie, the local butcher's son, slipped up behind the calf and wrung its tail. It soon took off and my brother did too…. These are a few things I remember about the rural school fair."[3]

Many readers, both old and young, can relate to these fair experiences or recall their own with even as much clarity!

The school-fair concept soon spread all over Canada. Agricultural educators across the country recognized the benefits of school fairs. Each province set up some sort of system whereby schoolchildren participated in a local school fair. The school fair enjoyed its heyday in the 1920s. Then the Great Depression took a terrible toll on the numbers. During World War II, government grants were discontinued. This led to the end of school fairs as separate entities under the Department of Agriculture. Agricultural fairs stepped into the void and assumed sponsorship of the school fairs in many districts, or at least offered special classes for children in their exhibit halls.

The school fair was designed to be the agricultural fair in miniature. Educators, teachers, parents and schoolchildren all recognized the value of agricultural fairs, and squeezed all the fun and benefit into a new little cousin known as the school fair. No wonder the school fair became soundly entrenched in local Canadian society.

School parade, Middlesex County, Ontario. Every year, each school fair had a different theme. The "High Cost of Living" was the chosen theme for this 1920 parade. The school kids dressed the theme?

School fair, Alberta, early 1900s. School fairs were for schoolchildren only. Grain, vegetables and chickens are lined up for judging. School fairs were popular until World War II, when they merged with agricultural fairs as a wartime measure. The basic idea behind the school fair was to encourage progressive agricultural techniques and improvements to the rural lifestyle through competition. To accomplish this goal, school-fair organizers copied the methods used by the highly successful agricultural fairs but adapted the scheme to school-kids only. It was also hoped that these children would grow up to become more active supporters of the local fair; once they became hooked at the school fair, they would support fairs for the rest of their lives.

4-H

The 4-H movement was started in the U.S. around 1900. Clubs are now found all around the world. The first clubs were organized in Ontario and Manitoba in 1912. They were originally called Boys and Girls Clubs. The clubs were created by progressive-minded individuals to promote farming and rural lifestyles and, later, homemaking skills. The first clubs were primarily concerned with animal husbandry and gardening. Over the years, the 4-H repertoire has expanded greatly to a wide spectrum of interests. In 1931, a national organization was established to coordinate the clubs across Canada. In 1952, the name 4-H was adopted. The four H's stand for Heart, Head, Hands and Health. They are open to youth under the age of twenty-one. Each club is led by one or more adult instructors. Youths are required to complete at least one project each year. Members' projects, or achievements, are shown to the public, and where else but the local fair. Agricultural societies have always been avid sponsors of Boys and Girls clubs and their successors, the 4-H clubs. In 1953, there was a total of 587 4-H clubs in Ontario, and 342 were sponsored by agricultural societies. Fairs and 4-H have become inseparable. Currently, there are 1,300 leaders and 4,800 4-H members across Canada.[4]

Boys and Girls Camps

A cousin of the 4-H movement was the Boys and Girls Clubs. These clubs were designed to introduce youth to progressive farming techniques and rural lifestyles. They differed from 4-H in their total emphasis on agriculture and the subjects taught in each club.

A logical offshoot from clubs was the organization of camps and integrated short courses for the educated rural youth. Instead of offering a weekly club, it was reasoned, a more intensive camp session would be

Farm boys proudly posing with their best grain exhibits. Children and fairs have always been a perfect fit. Notice the lack of shoes!

more useful. These boys and girls camps became especially popular across the Prairies. They involved intense courses, usually a few weeks long, designed to educate rural youth in an assortment of topics. Professional teachers and organizers were provided by the Ministry of Agriculture and regional universities. The obvious site for these camps was the local fairgrounds, which had buildings and facilities and excellent volunteer instructors. The camps were immensely successful. One held at Brandon fairgrounds in 1947 had over a thousand students enrolled.

Fairs and the boys and girls camps went hand in hand. The Saskatoon Fair, for instance, gave Hereford calves to farm children of local clubs in the spring of 1929. The children were to feed and care for the calves over the summer and return the calves for auction in the fall. In cooperation with the local board of trade, 1,060 calves were distributed and over 1,000 were returned for the auction!

Junior Home Economics Club canning competition. The three teams pictured are competing against one another. 4-H and homemaker clubs used the local fair to display the skills they learned in their clubs.

Youth Promotions

Fairs have always tried to encourage children to attend and participate. A legion of schemes have been used to attract kids to the fair. Some have been very successful. Many fairs have special children's days with reduced or free admission for children, coupled with special entertainment. In recent decades, a whole section of children's entertainers have grown up among the fairs. Everything from musicians to puppets to clowns to petting zoos to face painting have been used to make the children's visit to the fair enjoyable. The midway companies have also been active in this area. While midways were always youth-oriented, special Kiddyland sections are now organized for the smaller fry. Kiddyland has become immensely popular with children and parents alike. Midway companies have also initiated discount rates and special days to attract the little fair-goers and their parents. Pay-one-price days have always been popular on midways. In 1952, kids could ride all day for 5¢. Today it costs …well, a little more.

Children ride the motorcycles, Central Canada Exhibition, Ottawa, 1950s.

Various gimmicks used over the years to attract children to the fair have been less successful. In 1952, The Medicine Hat Fair decided to give out a dollar bill to every kid in the downtown core to spend at the fair. Unfortunately, the scheme was too well advertised and hordes of munchkins were waiting when the fair directors pulled up to the designated location in their car. Anxious kids besieged the directors so thoroughly, they couldn't open the car door. It took dozens of police and

Home Garden Contest winner Dorland Wood from Selby, Ontario. To promote agriculture among youth, the Ontario Ministry of Agriculture sponsored home garden contests. The winner stands stiffly at attention in his prize-winning vegetable patch.

187

Duncan Gray of Temiskaming District in Northern Ontario poses proudly with his winning sheaf of wheat. The badge reads "4th Prize Ontario School Fairs." Northern Ontario was proud of its fairs and its ability to go head-to-head with Southern Ontario.

firemen to rescue the directors and solve the ensuing traffic snarl. The scheme was too successful and had to be scrapped. The Saskatoon Fair was slightly more successful with its prize draws. Every child attending on Kids' Day was eligible for a draw featuring a Shetland pony. More than 5,000 kids watched in hushed awe as a winning name was drawn in the first-ever pony draw. The little boy whose name was drawn shyly approached the pony, nervously reached into his pocket and produced two sugar cubes, which he slowly offered to his new pony. The donor of the pony broke into tears and vowed he would donate THREE Shetland Ponies the next year. And he did!

Another successful method of attracting children to the fair was focusing on the school holiday. In many communities, schoolchildren were given a day off, or part of one, to attend the fair. Of course, special attractions and discounts were put in place on that day. These included children's sports, often organized by specific groups such as the YMCA. A 1925 fair list of boys' events included a sack race, three-legged race, wheelbarrow race, half-mile race and school relay race. Girls had different events, including a potato race, pea and spoon race, hobble skirts race, 75-yard dash and shuttle relay. Clearly, the boys were not yet running with the girls. Prizes were the norm in these competitions, not cash. Prizes could be solicited from local businesses, thus sparing the fair any expense. A sample of such prizes was as follows:

1st—rugby ball	fielder's glove	ring
2nd—watch	baseball bat	umbrella
3rd—flashlight	1st baseman's glove	perfume[5]

A Day At the Fair: A Young Boy Remembers

How much did the local fair mean to rural youth? The following is an account from the year 1910 of one farm boy's day at the fair.

188

"Those who did not experience it could never know the thrill the farm boy felt on fair day. It was 'the' day of all the year and neither Christmas Day, birthday nor election day could compare with it. It presented more new experiences, more strange places, more funny smells, more dazzling sights and more new faces than the farm boy would encounter in an average half year. Going back to the humdrum duties of driving horses and milking cows and grubbing roots after the fair was not easy; it was like eating dried bread after a feast of raisin pie.

"With cattle, horses, pigs and baking powder biscuits to be delivered and shown, it was a busy day as well as an interesting one. That old joy-killer, the alarm clock, sounded at 4:00 A.M. on fair day and the voice of the elder MacEwan boomed through the house, reminding those sufficiently awake to comprehend that no time could be lost because cattle and pigs must be delivered to the grounds before the heat of the mid-morning sun.

"The best set of harness had been carefully polished the day before, but there were a hundred jobs that could not be done in advance of departure hour. Overalls were pulled on to furnish some superficial protection to Sunday clothes, and with full knowledge that shoes would not remain clean, they were polished with special care. Then pigs were crated and loaded; feed was bagged and piled above the pig crates and unwilling cows were haltered and tied behind the wagons.

"The show horses were hitched and driven, or tied to the hitching rings of the wagon team, and presented no special problem. The pigs, trying to turn in the crates, usually jackknifed and made blasphemous squeals, but these were as nothing compared to the noises we would hear later in the day.

Junior Red Cross Exhibit, Calgary Fair, 1922. When the Red Cross had a message to deliver, the place to deliver it was at the local fair. This display certainly fits under "E" for education. The children look quite stunned. Maybe they were not used to posing for the photographer, or maybe they were just overwhelmed by the whole fair experience.

The Bug House, early 1900s. For a dime, the kids could venture into the wonderful world of insects, from a midway point of view.

"Cattle presented most of the problems. Without exception they refused to cooperate. They did not want to go to the fair. When the wagon started to move, they braced themselves, lay down and bellowed muffled but nasty sentiments and when they found it possible, broke their halter shanks and bounded to the most remote side of the pasture.

"The first exhibitors to arrive at the fair grounds claimed the best stalls or tied to the best trees. Judging was scheduled to begin at 10:00 A.M. but never started before 11. Excitement was mounting. Amateur showmen pleaded with disgruntled cows to look their best, at least to stand still long enough for the judge to recognize some semblance of domestication in the alleged breed.

"The most painful defeats were the ones near home. It was difficult to believe that any other herd bull in the district could or should be placed above our Glenwillow Romeo. And with all the neighbours watching, it was not easy to accept less than a red ribbon and do it with grace. There had been a mother's warning, however, that sportsmanship was more important than victory and it was a good test in discipline. And if one did not get a reasonably good share of first prizes, there was always some kind friend who could temporize with the thought that 'the judge is of the white-collared variety and must be pardoned for a certain number of mistakes.'

"Size in livestock and other things, ruled in those years. None but big potatoes could command any respect and always there was a pumpkin or a squash that would fill a bushel basket and was said to be 'the biggest in the world.' We knew we did not have the biggest fair but it was nice to fix upon some distinction and who was to dispute the claim that we had the biggest pumpkin of all time?

"The horse ring was really the centre of attraction. Unruly colts, spirited drafts, glamorous drivers and obstreperous stallions added their peculiar touches and there were those families who ate their picnic lunches there beside the horse ring so they wouldn't lose their places when judging was resumed for the afternoon. Informed horsemen were supposed to be able to name the sires of the best entries, and the colts by Flash Baron, by Baron's Pride, were supposed to have a congenital advantage in the draft classes.

"While the horse show held the spotlight, no sane person could have believed that at the greatly expanded and matured fair, held on the same grounds in 1949, only six head of draft horses would be shown.

"We sometimes exhibited seed grain but in wheat there was one sample, shown repeatedly for five years, that was considered unbeatable. The bachelor owner with nothing else to do, had spent all one winter hand-picking it and there was no rule to prevent him from bringing it back year after year to repeat his championship. Such was not uncommon in the ladies' work department. The cookies and cakes were supposed to belong to a contemporary generation, but in fancy work, there was no proof that some of the pieces had not won at the World's Fair at Paris (1798). One farmer from the North whose wife had been dead long since, remembered her at least once a year by exhibiting her hooked mat. It was invariably good for a red card and a dollar prize, until moths sabotaged the masterpiece.

"Judging having been completed, the squealing pigs, lowing cattle and whinnying horses were silenced temporarily with feed. Then, leaving their elders to argue about the judges placings, boys shed their overalls, whisked the less tenacious dirt from their shoes with a handful of hay and disappeared to lose themselves in less familiar but more exciting departments of the fair.

"Machinery row held some attractions, and we had to see the ball game, especially if Tom Redmond was pitching. Neither the grand champion Clydesdale stallion nor the imported judges from the University

Educational display promoting children's health. In this case, baby care is the topic. Fairs were ideal places to reach a large segment of Canadian society with your message.

could command as much youthful admiration as was breathed for that hero of the baseball diamond.

"There were side shows, and if the farm boy had to choose between a two-headed calf show and a two-legged girl show, his agricultural interests were easily restrained. If the decision was not an easy one, it was because the girl show cost 15¢ while the calf monstrosities could be seen for free." [10]

Deep in the lad's thoughts that day was the hope that he would meet up with one of the neighbourhood daughters or hired girls. Sooner or later, the meeting was sure to take place, much to the relief of both parties. Courting was part of the fair and sometimes the farm boy was missing when it came time to start home with the cattle.

There were so many things that a boy and girl could do together; they could visit the fortune teller; they could have their pictures taken as they held hands and sat on the thin end of the new moon or they could buy a ride on something resembling a merry-go-round, which at best was about as thrilling as a fast ride on an empty hay rack. They could buy peanuts, lollipops, popcorn or a mysterious brand of candy floss that resembled a webworm's cocoon but tasted somewhat better.

Everybody attending a fair in those days went home with some treasure, bought at half its value. The novelty specialist offered an invention for peeling potatoes at double the time-honoured speed and a longhaired "professor sold a cure for baldness," or a bottled medicine guaranteed to have the same affect upon the human body as a rebore job and general overhaul on the old car. The local wholesaler in drugs betrayed the medi-

Souris, Manitoba, Clothing Club, 1943. This 4-H club has won the T. Eaton trophy for their display at the Brandon Fair. The T. Eaton Company recognized the importance of fairs in Canadian culture. They regularly donated trophies and prizes to agricultural fairs all across Canada. These symbols of accomplishment are among the prized possessions of many fair exhibitors past and present.

cine man's secret, however, and the word spread quickly that any qualities of magic in the health-giving concoction were due to either the Epsom salts, the colouring matter or the flavouring that went into the water, because there was nothing else.

For the young man, nothing presented greater possibilities than the pocket-sized gadgets supposed to possess wonders far beyond those of a submarine's periscope. Placed at a keyhole, it should reveal everything on the inside, to the eye of the owner. It appeared cheap enough at 50¢, but as time was to prove, the money would have been better invested in peanuts.

For the girls, there was a perfume, promising husband, home and happiness. The fur trader had employed scents to lure his victims into traps; why should the girls overlook the possibilities? The noisy salesman called his product "matrimony assured," or something like that, and made what seemed a most reasonable offer in distributing those bottles of dollar value: "Pay 50¢ now and mail the balance to the manufacturer in Toronto, after you get a husband."

Beauty and a Beast, and some good-looking fair queen contestants too! Four contestants pose with a bovine beauty at the 1927 Pacific National Exhibition in Vancouver. Clothing and hairstyles have changed somewhat over the years, but many fairs still hold similar contests. The girls look remarkably comfortable atop their perch.

A similar contest to the one pictured on the previous page from 1954. Styles have certainly changed, and the bull is gone. Regional contests have been utilized to provide a contestant pool.

"The fair offered a wide variety of experiences. Although they could not all be good, experience is a good teacher and some of us learned that in spending our money, we should depend upon our own judgment more than upon the claims of the transient salesman. The person who paid 50¢ for a keyhole periscope that wouldn't work was less likely to buy a city property at Aklavik.

"When the excitement of fair day was subsiding, we had the cattle and other exhibits to take home. The animals were tired of it all and went back to the home pastures with more speed and alacrity than they displayed on the outward journey. Otherwise, for tired and weary travellers, Napoleon's Retreat from Moscow bore a striking similarity as we approached the end of a 19-hour day. Swollen and blistered feet were protesting too many hours in newish boots normally reserved for church attendance. It was never very easy to get "$1.98" dress boots that had more than a suggestion of resemblance to the shape and size of my feet anyway.

"Remorse in some form could not help following such a mixture of pleasures. Sore feet was but one of the penalties; indigestion might be another, and when the silver coins remaining in the little square purse were counted, there were some soul-searching questions about the wisdom of it all. When at the MacEwan breakfast on a morning following a fair, I confessed that my escapades had involved the spending of 90¢, I was told by an earnest parent that if I thought I was making an impression upon a certain feminine heart, it was time I realized that no really good girl would approve of such reckless spending.

"But there was much that was good arising from fair day. It was a welcome and needed change and sometimes it was the only deviation from monotonous farm work that summer afforded. Coupled with the peanuts, the fortune telling, the neighbour girls and the candy floss, there was experience in realities that led to a change of viewpoint about many things. The herd bull appeared in a new light and invariably there was a pronouncement that the next one must be a better one. We had seen mowing machines of different makes standing side by side and made up our minds about which would suit us best. We saw new varieties of seeds and vegetables and had proof that certain fruits would grow on Carrot River Valley soil. We saw what neighbors were doing and compared notes with them. Unless he turned its back upon it all deliberately, impressionable youth could not help benefiting from that annual outing."[6]

The author of this selection, Grant MacEwan, went on to become a University professor, author and prominent Canadian. And yes, he was a lover of fairs.

Today's fairs still promote children's activities very vigorously, and most fairs are determined to make a child's experience at the fair a happy one. There are not many places in today's world where every member of the family, from toddlers to grandparents, can find such a variety of sights and entertainment as the local fair. The event appeals to all ages by its very diversity. And entertaining and educating children is a top priority of every fair. If a fair is to build a good reputation within its community, that reputation should start with the young. Ah, those memories of being young at the fair!

Fairs are about tradition. They often feature many generations of the same family making the fair "happen." These two photos were submitted by a proud mother. They feature mom posing with her first-prize scarecrow in 1957, and her daughter's first-prize entry in 2001. She writes: "We both won first at our local fair, both of us were equally excited about our red ribbon (we cannot remember prize money), both fairs took place the same week, and we were both the same age!" What more needs to be said.

FARM BOY/GIRL CAMPS: RURAL SUMMER SCHOOL

Farm boy/girl camps were an important agricultural-education feature throughout Canada in the first half of the twentieth century. They were basically short educational courses designed to improve farming practices and rural lifestyles. The idea behind instructing boys and girls was to teach the next generation proper agricultural practices. It was easier to start out right with youth than convert or change the older generation. The increasingly active role of women in rural development was credited with kick-starting junior groups like 4-H, school fairs and farm boy/girl camps. As a committee of agricultural society delegates and government

Farm Boys Camp, Saskatoon. A popular event as evidenced by the huge turnout. In the background lurks the midway sign. Education and entertainment, the twin devils of the fair industry.

agricultural representatives in Saskatchewan put it: "Time spent in counselling and advising adults, that are not desirous of counsel or advice, is time wasted, and that the only hope of accomplishing aught of permanent worth is restricted to directing the energies and formulating the ideas and ambitions of those... who are to become the adult population of tomorrow."[1]

The first farm boy camp was held in Winnipeg in 1913. Enrollment was limited to 104 boys. Over 370 applied, and the winners were selected by judging an essay on "The Prairie Home." The main object of these summer schools was education and agricultural improvement, but they also served as a holiday, introduction to camp life, fellowship meetings and a welcome change of pace for farm boys. While in Winnipeg, the boys toured the agricultural college, the stockyards and that most venerable of Canadian institutions, the T. Eaton store.

A typical farm boy camp in 1917 had the following schedule: the 200 students were kept busy from 6:30 A.M. to 11:00 P.M. They studied the best breed of cow for milking, best beef types and how to raise pigs for bacon. Some scientific topics included: care of farm animals, control of noxious weeds, proper killing and dressing of poultry for market, preparation of wool for shipping and summer fallowing of land. Team judging contests, hygiene classes, baseball, swimming and other sports were added for variety and entertainment; for their motto was: "All work and no play makes Jack a city boy."[2]

In 1926, a farm boy camp was established at the Cumberland County, Nova Scotia, fairgrounds. One director expressed his opinion of the camp's results:

"I think these club members did much to improve the livestock part of the show. They procured good fitting halters and trimmed up their livestock. Previous to this, some of the showmen were rather careless and didn't have their animals trained to lead and would come into the ring leading an animal with rope around its neck. Needless to say, they didn't have much control and didn't add much to the show."[3]

The agricultural societies were very prominent in the farm camp program. They supplied facilities in their fairgrounds, expert instructors and often financial support. In most cases, the agricultural society was the sponsoring organization.

THE GREAT DEPRESSION

One of the darkest episodes in Canadian history was the Great Depression of the 1930s. The world economy collapsed and many Canadians were thrown out of work. Western Canada suffered through a tremendous drought, and commodity prices collapsed. Canadians had fewer dollars to spend; and every corner of society was affected. Fairs were not immune to the economic downturn. Attendance and spending dropped at fairs, simply because people did not have the cash to spend. Faced with declining attendance, fairs reacted in various ways. Admission fees were reduced, often substantially. Many fairs reduced their admissions from 50¢ to 25¢, while others, faced with non-existent crowds, gave free admission. Saskatoon Fair, in the heart of the drought-stricken Prairies, attracted 120,000 patrons with free admission in 1932. Children were given free admission at many fairs, while others admitted the unemployed free. Many agricultural societies faced financial disaster because they had borrowed heavily to upgrade their facilities and grounds while others ran deficits as revenues shrank. Societies unable to pay mortgages lost their fairgrounds. It was a time for soul-searching in many communities. Norfolk County in Ontario advanced the cash to keep Simcoe Fair afloat. The city of Regina assumed $125,000 in debt racked up by the Regina Fair, and the fair kept going. The government of Saskatchewan increased grant amounts to encourage fairs to continue. In 1930, 90 percent of Canadian class A fairs lost money. In 1933, the government of Manitoba cancelled grants to C and D fairs, with the resulting loss of thirty-six agricultural societies. The departmental-judge system and field-crop competitions were cancelled, albeit briefly, to cut costs.

The number of fairs was dramatically reduced during the 1930s. The farm population was hit hard by the bad economic times and the number of farmers and farms fell dramatically. Hard on the heels of the demise of farm communities came the disappearance of many fairs. The 1920s had been the high water mark as far as the number of agricultural fairs across Canada was concerned. But they declined from then as follows, in selected provinces:

Province	1928	1935	1945
Alberta	63	18	-
Manitoba	82	25	54
Ontario	354	307	225
Saskatchewan	146	29	47

The Prairie provinces suffered tremendously as drought added misery to tough economic conditions. Many fair societies disappeared completely in the dust bowl areas of the Prairies, while others hibernated during the 1930s, only to emerge later. Mergers between agricultural societies were also common, but then many people believed there were too many fairs in some provinces to begin with. To them, mergers or rationalizations were a good thing because fewer and stronger fairs would be of benefit to the whole agricultural-fair network. It was hard for a community to give up its fair. Community prestige was lost, and community spirit suffered.

But it was not all doom and gloom. As the old adage goes, "When the going gets tough, the tough get going." The fairs that did survive tried harder. They abandoned expensive stage shows for cheaper, more creative ones. Saskatoon Fair sponsored a class for Bennett Buggies: a car pulled by a team of horses because the owner could not afford gasoline. Weyburn Fair in Saskatchewan produced a lineup including a school parade, football, baseball, chariot races, auto collisions (obviously a forerunner to demolition derbies!), a yo-yo contest and a public wedding ceremony! When one fair cut the purse for horse racing from $2,000 to $175, each director was instructed to bring one racehorse to the fair. Failure to produce a horse on fair day led to a $2 fine!

Some fairs actually flourished during the Great Deppression. Lacking money for expensive diversions, fairs suddenly became good entertainment value to the public. Cheap entertainment was the order of the day, and fairs certainly fit this bill. Where else could you go for a day's outing for 25¢? While many weaker fairs declined or disappeared, stronger and more aggressive organizations actually managed to increase attendance during the 1930s. Well-managed fairs came through the Depression with flying colours. So successful were some fairs, in 1931 the Dominion government threatened to suspend grants to fairs that were making money. What a strange twist of fate: penalizing success! Nevertheless, Canada and its fairs survived the Great Depression. They were both a bit bruised and battered, but they survived.

Brandon Fair, 1934.

Conclusion

AND THERE YOU HAVE IT: two hundred and fifty years of Canadian fair tradition in a nutshell. Time and space prohibit me from paying justice to this topic. Agricultural fairs are so diverse, so multi-faceted, such an immense subject, it is impossible to cover the topic in one book. They contain much that cannot be experienced in the pages of a book or on a computer screen. The sights, sounds and smells that flood from the midway, or the livestock barns or the homecraft exhibits or the commercial displays or the parade or the entertainment must be personally experienced to be fully enjoyed. To truly appreciate a fair, it is necessary to be there in person.

Canadian agricultural fairs are still a part of Canadian culture. They are found all across Canada. They mirror Canadian history, geography, economics and culture. They are community based, with the fairgrounds often doubling as the community centre. They are run by volunteers. And above all else fairs are fun and good times. They appeal to mankind's spirit of social interaction.

What does the future hold for Canadian fairs? Today's world moves at a rapidly changing pace. Modern technology has radically changed the face of our society. Many old and venerable institutions have been swept away. Others have been changed beyond recognition. But fairs will still be there in the twenty-first century. Some will fail, but others will flourish. They are traditional, educational and adaptable. Yet, if they lose touch with their roots, fairs will falter. It is a fine line fairs walk between tradition and the space age. Fairs will continue as long as the community spirit remains strong. The community may be a rural township or a large city, but it takes a strong community effort to make the fair successful.

I hope this book has left the reader with an appreciation for the traditions embodied by Canadian agricultural fairs and the role they have played in Canada's history. I also hope this work will inspire Canadians to support fairs and their communities in any way they can to keep this venerable Canadian institution not just alive, but growing and flourishing. I hope I have inspired some readers to look at the fairs in a new light. And never forget the immortal words of that small boy who, when asked what he liked best about the fair, replied: "It makes everybody happy."

Fair Listings

Ontario

May   Schomberg Ag. Soc.

June   Aurora Ag. Society Fair & Horse Show
            Brooklin Ag. Soc.
            Clinton-Huron Central Ag. Soc.
            Gloucester Fair
            Hensall Ag. Soc.
            Leamington Ag. Soc.
            Maxville Ag. Soc.
            Millbrook Ag. Soc.
            Strathroy Ag. Soc.
            Warren Ag. Soc.

July   Avonmore Ag. Soc.
            Beachburn Ag. Soc.
            Delta Ag. Soc.
            Dresden Ag. Soc.
            Kenora Ag. Soc.
            Lakefield Ag. Soc.
            Lansdowne Ag. Soc.
            Listowel Ag. Soc.
            Peterborough Ag. Soc.
            St. Mary's Ag. Soc.
            Tweed Ag. Soc.
            Zurich Ag. Soc.

Aug.   Arnprior Ag. Soc.
            Aylmer Ag. Soc.
            Barrie Fair & National Horse Show
            Bayfield Ag. Soc.
            Blackstock Ag. Soc.
            Bonfield Ag. Soc.
            Burkes Falls Ag. Soc.
            Campbellford Ag. Soc.

Canadian Lakehead Exhibition
Canadian National Exhibition
Carden Ag. Soc.
Central Canada Exhibition (Ottawa)
Centreville Ag. Soc.
Charlton Ag. Soc.
Chesterville Ag. Soc.
Cobden Ag. Soc.
Cochrane Ag. Soc.
Coe Hill Ag. Soc.
Comber Ag. Soc.
Desboro Ag. Soc.
Drayton Ag. Soc.
Dryden & District Ag. Soc.
Dunchurch Ag. Soc.
Dungannon Ag. Soc.
Dunnville Ag. Soc.
Durham Ag. Soc.
Elmira Ag. Soc.
Emo Ag. Soc.
Emsdale Ag. Soc.
Fenelon Ag. Soc.
Foley Ag. Soc.
Haliburton County Exhibition (Minden)
Harrow Ag. Soc.
Hymers Ag. Soc.
Kincardine Ag. Soc.
Kinmount Ag. Soc.
Lombardy Ag. Soc.
Maberly Ag. Soc.
Magnetawan Ag. Soc.
Manitowaning Ag. Soc.
Markdale Ag. Soc.
Marmora Ag. Soc.
Massey Ag. Soc.
Matheson & District Ag. Soc.

Meaford & St. Vincent Ag. Soc.
Melbourn Ag. Soc.
Merrickville Ag. Soc.
Mitchell Ag. Soc.
Mount Forest Ag. Soc.
Murillo Ag. Soc.
Napanee Ag. Soc.
Navan Ag. Soc.
North Shore Ag. Soc.
Odessa Ag. Soc.
Orangeville Ag. Soc.
Palmerston Ag. Soc.
Parham Ag. Soc.
Paris Ag. Soc.
Perth Ag. Soc.
Porquic Ag. Soc.
Port Perry Ag. Soc.
Powassan Ag. Soc.
Providence Bay Ag. Soc.
Quinte Ex. (Belleville)
Rainy River Valley Ag. Soc.
Riceville Ag. Soc.
Ridgetown Ag. Soc.
Rosseau Ag. Soc.
Shannonville Ag. Soc.
Shedden Ag. Soc.
Smithville Ag. Soc.
South Mountain Ag. Soc.
Stirling Ag. Soc.
Stormont County Ag. Soc.
Sutton Ag. Soc.
Teeswater-Culross Ag. Soc.
Tillsonburg Tri-County Fair
Trout Creek Ag. Soc.
Vankleek Hill Ag. Soc.
Wilberforce Ag. Soc.
Williamstown Ag. Soc.
Woodstock Ag. Soc.

Sept.   Aberfoyle Ag. Soc.
Acton Ag. Soc.

Almonte Ag. Soc.
Ancaster Ag. Soc.
Arran-Tara Ag. Soc.
Arthur Ag. Soc.
Beaver Valley Ag. Soc.
Beaverton Ag. Soc.
Beeton Ag. Soc.
Binbrook Ag. Soc.
Bobcaygeon Ag. Soc.
Bolton Ag. Soc.
Bracebridge Ag. Soc.
Brampton Ag. Soc.
Brooke, Alvinston & Watford Ag. Soc.
Bruce Mines Ag. Soc.
Brussels Ag. Soc.
Caledon Ag. Soc.
Caledonia Ag. Soc.
Cambridge Ag. Soc.
Carp Ag. Soc.
Chatsworth Ag. Soc.
Chesley Ag. Soc.
Coldwater & District Ag. Soc.
Cookston Ag. Soc.
Donnybrook Ag. Soc.
Drumbo Ag. Soc.
Dundalk Ag. Soc.
Durham Central (Orono) Ag. Soc.
Embro-Zorra Ag. Soc.
Englehart Ag. Soc.
Exeter Ag. Soc.
Fergus Ag. Soc.
Feversham Ag. Soc.
Forest Ag. Soc.
Georgetown Ag. Soc.
Glencoe Ag. Soc.
Grand Valley Ag. Soc.
Great Norhtern Ex (Collingwood)
Hanover Ag. Soc.
Harriston Minto Ag. Soc.
Highgate Ag. Soc.
Houghton Ag. Soc.

Huntsville Ag. Soc.
Ilderton Ag. Soc.
Iron Bridge Ag. Soc.
Kingston Ag. Soc.
Kirkton Ag. Soc.
Langton Ag. Soc.
Lincoln County Ag. Soc.
Lindsay Ag. Soc.
Lucknow Ag. Soc.
Madoc Ag. Soc.
McDonald's Corners Ag. Soc.
McKellar Ag. Soc.
Middleville Ag. Soc.
Mildmay-Carrick Ag. Soc.
Milton Ag. Soc.
Milverton Ag. Soc.
Mohawk Fair
Neustadt Ag. Soc.
New Hamburg Ag. Soc.
New Liskeard Ag. Soc.
Niagara Regional Ag. Soc.
Noelville Ag. Soc.
Oakwood Ag. Soc.
Orillia Ag. Soc.
Oro Ag. Soc.
Owen Sound Ag. Soc.
Paisley Ag. Soc.
Parkhill Ag. Soc.
Petrolia & Enniskillen Ag. Soc.
Picton Ag. Soc.
Plympton & Wyoming Ag. Soc.
Porcupine Ag. Soc.
Port Hope Ag. Soc.
Ramona Ag. Soc.
Renfrew Ag. Soc.
Richmond Ag. Soc.
Ripley-Huron Ag. Soc.
Rocklyn Ag. Soc.
Rodney-Aldborough Ag. Soc.
Russell Ag. Soc.
Seaforth Ag. Soc.

Severn Bridge Ag. Soc.
Shelburne & District Ag. Soc.
Six Nations Ag. Soc.
South River-Machar Ag. Soc.
Spencerville Ag. Soc.
Stayner Ag. Soc.
Stisted Ag. Soc.
Stratford Fair
Strong Ag. Soc.
Sunderland Ag. Soc.
Sydenham Ag. Soc.
Tavistock Ag. Soc.
Thorndale Ag. Soc.
Uxbridge Ag. Soc.
Wainfleet Ag. Soc.
Wallacetown Ag. Soc.
Warkworth Ag. Soc.
Wellesley Ag. Soc.
Western Fair Assoc. (London)
Wiarton Ag. Soc.

Oct.   Burford Ag. Soc.
       Dorchester Ag. Soc.
       Elmvale Ag. Soc.
       Erin Ag. Soc.
       Howick-Turnberry Ag. Soc.
       Markham Fair
       Metcalfe Ag. Soc.
       Moore Ag. Soc. (Brigden Fair)
       Norfolk County Fair (Simcoe)
       Norwood Ag. Soc.
       Rockton Ag. Soc.
       Roseneath Ag. Soc.
       Tiverton Ag. Soc.
       Walkerton Ag. Soc.
       Woodbridge Ag. Soc.

Nov.   Royal Ag. Winter Fair

Quebec

May
: Becancour Ag. Soc.
Temiscamingue Ag. Soc.

June
: Drummondville Exhibition (Matapedia)
Ormstown Exhibition
Richelieu Ag. Soc.
Rouville Ag. Soc.

July
: Argenteuil Ag. Soc. (Expo Lachute Fair)
Chicoutimi Ag. Soc.
Expo Agricole de Bellechasse-Dorchester Inc.
Exposition Agricole de St-Hyacinthe
Exposition Agricole du Centre du Québec (Trois-Rivières)
Exposition du Bassin de la Chaudière
Kamouraska Ag. Soc.
Lotbinière Ag. Soc.
Portneuf Ag. Soc.
Rimouski Ag. Soc.
Rive-Nord Ag. Soc.
Verchères Ag. Soc.

Aug.
: Abitibi Ag. Soc.
Brome County Ag. Soc.
Cookshire Ag. Soc.
Expo Agricole Commerciale (St. Félicien)
Expo Québec
Exposition Agricole de Beauce Inc
Huntingdon Ag. Soc.
Missisquoi Ag. Soc.
Montmagny Ag. Soc.
Pont-Château Ag. Soc.
Pontiac Ag. Soc. (Shawville)
Stanstead Ag. Soc.
Temiscouata Ag. Soc.
Victoriaville Ag. Soc.

Sept.
: Richmond Ag. Soc.

Saskatchewan

May
: Maple Creek Ag. Soc.
Vanscoy & District Ag. Soc.

June
: Alameda Ag. Soc.
Bengough Ag. Soc.
Biggar & District Ag. Soc.
Estevan Ex. Assoc.
Fairmede Ag. Soc.
Hanley & District Ag. Soc.
Kerrobert Ag. Soc.
Meadow Lake Ag. Soc.
Melville & District Agri-Park Assoc.
Moose Jaw Exhibition
Moosomin Ag. Soc.
Spiritwood Ag. Soc.
Swift Current Exhibition
Weyburn Ag. Soc.
Whitewood Ag. Soc.

July
: Abernathy Ag. Soc.
Central Butte & District Ag. Soc.
Coronach Ag. Soc.
Creelman Ag. Soc.
Glenavon Ag. Soc.
Lloydminster Exhibition
Maryfield Ag. Soc.
Melfort Ag. Soc.
Nipawin Ex. Assoc.
Ogema Ag. Soc.
Parkland Ag. Soc.
Perdue Ag. Soc.
Prince Albert Exhibition
Redvers Ag. Soc.
Regina Ex. Park
St. Walburg & Dist. Ag. Soc.
Stoughton Ag. Soc.
Yorkton Ex. Assoc.

Aug. Arcola Ag. Soc.
Battlefords Ag. Soc.
Golburn Ag. Soc.
Invermay Ag. Soc.
Kelvington Ag. Soc.
Nokomis Ag. Soc.
Radisson Ag. Soc.
Rosthern Ag. Soc.
Saltcoats Ag. Soc.
Saskatoon Prairieland Exhibition
Shand Ag. Soc.
Shaunavon-Admiral Ag. Soc.
Turtleford Ag. Soc.

Sept. Grenfell Ag. Soc.
Unity Ag. Soc.

Oct. Broadview Ag. Soc.
Denzil Ag. Soc.

Nov. Canadian Western Agribition Association (Regina)

Alberta

Mar. Grande Prairie Ag. Soc.

May   Blackfalds Days
Strathcona Country Classic

June Berwyn Fair
Chipman Fair Days
Hay Lakes Fair
Holden & District Ag. Soc.

July Big Valley Jamboree
Bowden Daze
Calgary Ex. & Stampede
Carstairs Rodeo
Edgerton Fair & Livestock Show

Edmonton's Klondike Days Exposition
Kinsella Bull-a-rama
Lamont & District Ag. Soc.
Medicine Hat Ex. & Stampede
Myrnam Fair & Fun Days
Strathmore & District Ag. Soc.
Summer Ssizzler Rodeo & Fair
Vegreville Ag. Soc.
Vermilion Ag. Soc. Fair
Westerner Days Fair & Exposition

Aug. Ashmont Heritage Days
Barrhead Ag. Soc. Blue Heron Fair
Bashaw Rodeo
Beaverlodge Ag. Fair
Bentley & District Ag/Elks Rodeo and Fair
Bezanson Ag. Society Fair & Horse Show
Boyle August Rodeo
Buffalo Fall Fair
Carmangay Ag. Soc.
Castor Fair
Cochrane Horse Trials
Coronation Town & County Fair
Darwell & District Ag. Fair
Didsbury Fair & Rodeo
Drayton Valley Bench Fair
Duchess & District Fall Fair
Eaglesham & District Fair
Fort Assiniboine Hamlet Hoedown & Rodeo
Gleichen Fair & Rodeo
Harmon Valley Fall Fair
Innisfree & District Fair & Horse Show
Jamboree Days (Picture Butte)
Marwayne Fall Fair
Millarville Racing & Ag. Soc.
Mountain View County Fair & Rodeo (Olds)
Mundare Agri-Daze
New Sarepta Fair
Peace River Fall Fair & Horseshow

Ponoka County Fair
Rich Valley Fair
Rochester & District Ag. Fair
Round Hill & District Bench Show
St. Paul Rodeo
Tees Long Ears Days
Tofield's Country Fair
Valleyview & District Ag. Soc. Fair & Rodeo
Vilna & District Ag. Soc.
Wetaskiwin Ag. Soc.– Pioneer Days
Whoop-up Days Summer Fair & Rodeo (Lethbridge)
Wildwood Fair

Sept. Chestermere County Fair

Oct. Milo Fall Fair

New Brunswick

July Albert County Fair
Woodstock Old Home Week

Aug. Atlantic National Exhibition
Exposition Regionale du Madwaska Inc.
Kent County Agriculture Fair
Kings County Ag. Fair
L' Exposition Regionale de St-Isidore
Miramichi Ag. Ex. Assoc
Port Elgin Ag. Fair
Westmorland County Fair

Sept. Fredericton Ex. Ltd.
Provincial Livestock Show
Queens County Fair
Stanley Fair

Nova Scotia

July South Shore Ex. (Bridgwater)
Western N.S. Ex. (Yarmouth)

Aug. Annapolis County Exhibition
Barrington Fair
Cape Breton County Exhibition
Cumberland County Exhibition
Digby County Exhibition
Eastern Nova Scotia Ex. (Antigonish)
Halifax County Ex. (Middle Musquodoboit)
New Ross Community Fair
Nova Scotia Provincial Ex. (Truro)
Shelburne County Ex.
Western Kings Fair (Tremont)

Sept. Hants County Ex. (Windsor)
Nova Scotia 4-H Show (Truro)
Pictou/North Colchester Ex.
Queens County Ex. (Caledonia)

Oct. Maritime Fall Fair (Halifax)

Prince Edward Island

July Crapaud Ex. Assoc
PEI Potato Blossom Festival
Prince County Exhibition

Aug. Kensington Harvest Festival
L' Exposition agricole et le Festival Acadien de la region Evangeline
PEI Plowing Match & Ag. Fair (Dundas)
PEI Provincial Ex.

Sept. 4-H Rural Youth Fair (Abram Village)
Eastern Kings Ex. (Souris)

British Columbia

May — Cloverdale Rodeo & Ex. Assoc.

July — Abbotsford Agri-fair

Aug. — Arrowsmith Ag. Assoc. (Coombs)
Bella Coola Fall Fair
Brackendale Fall Fair
Bulkley Valley Ex. (Smithers)
Chilliwack Exhibition
Cluculz Lake Fall Fair
Cobble Hill Fair
Comox Valley Ex. (Courtenay)
Dawson Creek & District Exhibition
Interior Provincial Exhibition
Kiskatinaw Fall Fair
Kootenay Lake Ag. Fair (Crawford Bay)
Lakes District Fair Assoc. (Burns Lake)
Mayne Island Fall Fair
Nechako Valley Ex. (Vanderhoof)
Nicola Valley Fall Fair (Merritt)
North & South Saanich Ag. Soc.
North Peace Fall Fair (Fort St. John)
North Thompson Fall Fair & Rodeo (Barriere)
Pacific National Exhibition
Pender Island Fall Fair
Prince George Exhibition
Quesnel Fall Fair
Robson Valley Fall Fair
Skeena Valley Fall Fair (Terrace)
Tlell Fall Fair
Vancouver Island Ex. (Nanaimo)

Sept. — Agassiz Fall Fair & Corn Festival
Aldergrove Ag. Assoc.
Alverni District Fall Fair
Ashcroft & District Fall Fair
Cowichan Ex. Soc. (Duncan)
Creston Valley Fall Fair
Fort Fraser Fall Fair
Grand Forks & Distrcit Fall Fair
Kamloops Provincial Winter Fair
Kootnay Country Fair (Cranbrook)
Lillooet & District Fall Fair
Luxton Fall Fair
Mount Waddington Regional Fall Fair
Pass Creek Regional Ex. Soc. (Castlegar)
Peachland Fall Fair
Powell River Fall Fair
Princeton & District Fall Fair
Rock Creek & Boundary Fair Assoc.
Rossland Fall Fair
Salt Spring Island Fall Fair
Summerland Fall Fair
Williams Lake Harvest Fair

Manitoba

Mar. — Manitoba Royal Winter Fair (Brandon)

May — South Interlake (Rockwood) Ag. Soc.

June — Boissevain Fair (Turtle Mountain)
Brandon Provincial Ex (Summer Fair)
Cypress River Fair
Dauphin Fair
Deloraine Fair
Foxwarren Fair
Killarney Fair
Lundar Fair
Melita Fair (Arthur, E.D.)
Miani Fair & Rodeo
Neepawa Fair (Beautiful Plains)
Ninette Fair (Pelican Lake)
North Norfolk/MacGregor Fair & Rodeo
Rapid City Fair & Rodeo
Red River Ex. (Winnipeg)
The Pas Fair (Opasquia)
Treherne Fair

July	Arborg Fair	Oct.	Roland Pumpkin Fair

July Arborg Fair
Carberry Fair
Carman Country Fair (Dufferin)
Crystal City Fair (Clearwater)
Dugald Fair (Springfield)
Elkhorn Fair
Gilbert Plains Fair & Rodeo
Glenboro Fair
Hamiota Fair
Harding Fair
Manitoba Stampede & Ex. (Morris)
Manitou Fair
McCreary Fair
Minnedose Fair
Oak Lake Fair
Oak River Fair
Plumas Fair
Portage Industrial Exhibition
Reston Fair (Pipestone-Albert)
Rivers Fair
Rossburn Fair & Racemeet
Selkirk Triple S Fair & Rodeo
Shoal Lake Fair
Souris/Glenwood Fair
Strathclair Fair
Swan River Northwest Roundup & Ex.
Virden Fair

Aug. Birtle Fair
Gladstone Fair & Rodeo
Grunthal Fair (Hanover)
Kelwood Fair
Roblin Fair & Rodeo
St. Pierre Fair
St. Vital Fair
Teulon Fair
Winkler Harvest Festival & Ex. (Stanley)

Oct. Roland Pumpkin Fair

Nov. Brandon Provincial Ex. (Livestock Expo)

Newfoundland

May Agrifoods & Garden Show (Cornerbrook)

July Peach Festival (Salmon Cove)
Strawberry Festival (Humber Valley)

Aug. Brigus Blueberry Festival
Farm Field Day (St. John's)
Labrador Bakeapple Festival (Forteau)

Sept. Green Bat Ag. Fair (Springdale)
Humber Valley Ex. (Deer Lake)
Port au Port Ag. Fall Fair
Trinity Conception Fall Fair (Harbour Grace)

Photo Credits

Abrreviations:
- NAC - National Archives of Canada
- GA - Glenbow Archives
- UBA - University of Brandon Archives
- CNE - Canadian National Exhibition
- PAO - Provincial Archives of Ontario
- PAS - Provincial Archives of Saskatchewan
- PAM - Provincial Archives of Manitoba
- BCA - British Columbia Archives
- WFS - World's Finest Shows (Conklin's Midway)
- OAAS - Ontario Association of Agricultural Societies
- CCE - Central Canada Exhibition (Ottawa)

Page
- v OAAS
- xii top: NAC, left: Bridgewater Fair (NS), right: Brandon Fair, Man.
- xiii CNE
- xiv-xv UBA
- 1 GA
- 2 NAC
- 4 WFS
- 5 GA
- 6 GA
- 7 GA
- 8 CNE
- 9 CNE
- 10 both GA
- 11 PAO
- 12 Sakatoon Exhibition
- 13 GA
- 15 OAAS
- 16 BCA
- 17 OAAS
- 18 GA
- 23 MacKennon
- 24 Nottingham Goose Fair Website
- 25 Sturbridge Fair
- 27 MacKennon
- 28 OAAS
- 30 OAAS
- 34 PAS
- 35 Brigden Fair, Ont.
- 38 GA
- 39 NAC
- 40 NAC
- 41 Simcoe Fair, Ont.
- 43 Kinmount Fair, Ont.
- 44 Western Fair (London), I. Sanmiya
- 46 NAC
- 47 top: Brandon Fair, bottom: Saskatchewan Public Library
- 48 Prince Albert Fair
- 49 GA
- 50 both: GA
- 51 both: GA
- 52 PNE (Vancouver)
- 53 top: GA bottom: NAC
- 54 both: BCA
- 55 CCE (Ottawa)
- 56 UBA
- 57 OAAS
- 58 GA
- 59 GA
- 60 top: UBA
- 60–63 GA
- 63 WFS
- 64 Mackennon
- 66 GA
- 68 top: GA bottom: MacKennon
- 69 GA
- 70 top: GA bottom: PNE
- 71 CNE
- 72 both: GA
- 73 top: OAAS, bottom: CNE
- 74 WFS
- 75 top: GA bottom: CNE
- 77 both: CNE
- 78 OAAS
- 79 PAS
- 80 CNE
- 81 CNE
- 82 WFS
- 83 GA
- 84 CCE
- 85 UBA
- 88 MacKennon
- 91 CNE
- 92 Armstrong Fair (BC)
- 93 Kinmount Fair
- 94 GA
- 95 both: GA
- 96 Markham Fair
- 97 GA
- 98 top: OAAS, bottom: GA
- 99 GA
- 100 CNE
- 101 CNE
- 102 OAAS
- 103 NAC
- 104 UBA
- 105 GA
- 106 GA
- 107 both: CNE
- 108 CNE
- 109 both: CNE
- 110 GA
- 111 GA
- 112 CNE
- 113 Western Fair, I. Sanmiya
- 114 CNE
- 115 UBA
- 119 Tom Bishop, 4-B Wild West Shows
- 120 PAM
- 121 NAC
- 122 top left: PAO, top right: PAO, bottom: NAC
- 123 top left: PAS, top right: PAS, bottom: UBA
- 124 top: NAC, middle: NAC, bottom: CNE
- 125 top left: GA, top right: NAC, bottom left: NAC, bottom right: NAC
- 126 top: CNE, bottom left: , bottom right: Sakatoon Fair
- 127 top left: NAC, top right: NAC, bottom: NAC
- 128 CNE
- 129 top: PNE, bottom: Peterborough Fair
- 130 Armstrong Fair BC
- 131 both: Peterborough Fair, Ont.
- 132 NAC
- 133 NAC
- 134 top: PAO, bottom: Bracebridge Fair, Ont.
- 135 GA
- 136 both: GA

137 top: Western Fair, London,
    I. Sanmiya
    bottom: NAC
138 CNE
139 top: GA, bottom: NAC
140 PAO
141 CNE
142 OAAS
143 top: CCE, bottom: CNE
144 GA
145 NAC
146 PAO
147 CNE
148 NAC
149 NAC
150 UBA
151 CNE
152 OAAS
155 NAC
156 NAC
157 NAC

158 Kinmount Fair, Ont.
159 Bridgewater Fair, NS
160 top: CNE, middle: OAAS,
    bottom: NAC
161 top: NAC, bottom: Port Perry
    Fair, Ont.
162 both: Kinmount Fair, Ont.
163 UBA
164 Bracebridge Fair, Ont.
165 Bracebridge Fair, Ont.
166 PAO
167 GA
169 GA
171 NAC
172 top: CNE, bottom: GA
173 top: Bobcaygeon Fair, Ont.,
    bottom: GA
174 top: GA, bottom left: GA,
    bottom right: GA
175 both: GA
176 both: GA

177 top: GA, bottom: CNE
179 Brandon Fair, Man.
180 PAO
181 OAAS
182 PAO
183 Williamstown Fair, Ont.
184 PAO
185 GA
186 top: OAAS, bottom: CNE
187 top: CCE, bottom: PAO
188 PAO
189 GA
190 GA
191 CNE
192 UBA
193 PNE
194 PNE
195 Anne Baptist, Brampton, Ont.
196 top: Sakatoon Fair,
    bottom: Prince Albert Fair
199 Brandon Fair

Endnotes

Fair Culture in Canada
1. Scott, Guy: A History of fairs in Ontario (John Deyell Ltd., Lindsay) p. 27

Fair Organizations
1. Coates, Kenneth and McGuinness Fred: Pride of the Land: An Affectionate Look at Brandons Exhibitions. (Pegus, Winnipeg 1985)
2. MacEwan, Grant: Between the Red and the Rockies. (University of Toronto Press, 1952)

The Fairgrounds
1. Scott, Guy: A History of Fairs in Ontario: (John Deyell Ltd., Lindsay) p. 58
2. Breen, David and Coates, Ken: The Pacific National Exhibition: An Illustrated History. (University of BC Press, Vancouver, 1982 p. 150

Ancient Traditions
1. Website, www.ihrinfo.ac.uk/cmh/gaz
2. Wood, p. 17
3. Walford, Cornelius: Fairs Past and Present. (A.M. Kelley, New York, 1968)
4. Wood p. 20-25
5. Walford p. 175
6. *see* Walford
7. Scott, Guy: A History of Fairs in Ontario. p. 8
8. Scott, p. 9
9. Website, www.ukfunfairs.com
10. Website, www.worldsfairs.com
11. Scott, p. ?

Come To the Fair
1. Melbourne, Australia Website www.IAFE.com

Plowing Matches
1. Underwood, Amber: Breaking Ground: Ont. Ploughman Assoc. (Ampersand Printing, 1987) p. 8
2. MacEwan, Grant: Between the Red and the Rockies. p. 43

Canada's Fair History
1. Rasmussen, K.: Trailblazers of Canadian Agriculture. (Agricultural Institute of Canada, 1985) p. 45
2. Rasmussen p. 7-24
3. Rasmussen p. 7-24
5. Rasmussen p. 95-123
6. County of Argentil History Page, Website on Quebec Fairs.

7. Scott p. 13-24
8. Reynolds, Nila. In Quest of Yesterday. (John Deyell, Lindsay, 1973) p. 98
9. Scott p. 58
10. For More on Prairie Fairs, *see* Grant MacEwan's Between the Red and the Rockies
11. *see* MacEwan, Grant. Between the Red and the Rockies. (University of Toronto Press, 1952)

Provincial Exhibitions
1. Hind, Henry Youle: Agricultural History of Canada. p. 45-47

The Midway
1. McKennon, Joe: A Pictorial History of the American Carnival. (Carnival Publishers of Sarasota, Florida, 1972) p. 16
2. McKennon, p. 16
3. McKennon, p. 42-56
4. McKennon, p. 129-130
5. Jones, Elwood: Winners: 150 Years of the Peterborough Exhibition. (Peterborough, 1995) p. 129-130
6. Scott, p. 127
7. Nelson, Derek: The American State Fair. (MBI Publishing, Ocala, Wisconsin, 1999) Chapter 7
8. Breen and Coates,
9. Nelson, p. 122-3
10. Nelson, p. 123
11. Nelson, p. 119
12. Nelson, p. 123
13. Nelson, p. 110
14. Jones, David: Midways, Judges and Smooth-Tongued Fakirs. (Western Producer Prairie Books, Saskatoon, 1983) p. 55
15. Jones, D., p. 55
16. McKennon, p. 130
17. Scott, p. 122
18. Scott, p. 127-133
19. Wilcox, George: A History of Exhibitions in Medicine Hat. (Friesen, Altona, Manitoba, 1996) p. 55
20. Jones, D., p. 58
21. Jones, D., p. 59
22. Jones, D., p. 63
23. Jones, D., p. 59
24. Jones, D., p. 59
25. Nelson, p.111

26 Nelson, p. 111
27 Nelson, p. 112
28 Coates and McGuinness, p. 20
29 Nelson, p. 112
30 Conklin Yearbook.
31 Conklin Yearbook
32 *see* CCE Website

Chicago World's Fair
1 Website www.chicagoworld'sfair.com

Entertainment
1 Claus, Louis; The Plate: A Royal Tradition. (Deneau Publishers, Toronto, 1984) p. 11
2 Scott, p. 99
3 Scott, p. 99
4 Grey, James: A Brand of Its Own: A History of the Calgary Stampede. p. 119
5 Wilcox, p. 52
6 Grey, p. 119
7 Jones, Elwood, p. 68
8 Cashman, Tony: The Edmonton Exhibition: The First Hundred Years. (Edmonton Ex Association, 1979)
9 Cashman
10 Various, Once Upon a Century: 100 Year History of the CNE. Robinson Publishing, Toronto, 1978) p. 119
11 MacEwan, Grant. Between the Red & Rockies. (University of Toronto Press, 1952)
12 Jones, D., p. 99
13 Coates & McGuinness, p. 145
14 Jones, E., p. 126-128
15 Coates & McGuinness, p. 145
16 Once Upon a Century, p. 51-69
17 Mowat, Ruth: All's Fair: A History of Williamstown Fair.
18 Once Upon a Century, p. 110
19 Nelson, p. 54-8
20 Breen, p. 122

Small Fairs vs Larger Fairs
1 Scott, p. 81-83
2 Jones, D., p. 104
3 Jones, D., p. 105

The Exhibit Hall
1 Jones, D., p. 98
2 Jones, D., p. 98
3 Jones, D., p. 98
4 Jones, D., p. 102

5 Jones, D., p. 102
6 Jones, D., p. 103
7 Jones, D., p. 101
8 Scott, p. 104

Livestock
1 Fleury, Kay.: Let's Go to the Fair: A History of the Weyburn Agricultural Society. p. 5
2 Jones, D., p. 119
3 Jones, D., p. 120
4 Scott, p. 84
5 Jones, D., p. 21
6 MacEwan, Grant. Between the Red & Rockies. (University of Toronto Press, 1952)
7 Stephenson, Bonnie: Agribition. (Centax Books, Regina, 1990) p. 68
8 Galbraith, A.: "40 Years Horse Judging Experience" Farm & Ranch Review, July 10, 1926, p. 8
9 Scott, p. 85
10 Jones, D., p. 24
11 Stephenson, p. 64
12 Frith, Joan: Treatise of a Society: A History of the Prince Albert Agricultural Association. (Friesen, Altona, Manitoba, 1983)
13 Sanmiya, Inge.: A Celebration of Excellence: A History of the Western Fair Association. (Western Fair, London, 2000)
14 Cashman, The Livestock King and The Beef Cattle Industry
15 Fish, Charles.: Blue Ribbons and Burlesque: A Celebration of Country Fairs.(Countryman Press, Woodstock, Vermont, 1998) p. 50-65
16 Millarville, Alberta Prize List, Stephenson, p. 67

Spring Fairs
1 Scott, p. 161

The Thrill of it All
1 Scott, p. 100-101
2 Scott, p. 101-102
3 Sunderland Fair, Ont. 150th Prize list.
4 Scott, p. 104
5 Jones, Elwood, p. 111
6 MacEwan, Agriculture on Parade

Farm Boy/Girl Camps
1 Jones, D., p. 100
2 Jones, D., p. 100
3 Cumberland Fair, N.S. Prize List.

Bibliography

Allwoods, John. *The Great Exhibitions*. Cassel, Collier, McMillan Limited, London, England.

Avery, Julie. *Agricultural Fairs in America*. Michigan State University, East Lansing, Michigan, 2000.

Breen, David and Kenneth Coates. *The Pacific National Exhibition: An Illustrated History*. University of British Columbia Press, Vancouver, 1982.

Campbell, Shirley. *The Interior Provincial Exhibition: Armstrong B.C.* Armstrong Advertiser, 1999.

Cashman, Tony. *Edmonton Exhibition: The First Hundred Years*. Edmonton Exhibition Association, Edmonton, 1979.

Clark, Cheryl and Brenda Vetter. *Reflections of an Era: A Look Back on 100 Years of the Battlefords Agricultural Society 1885-1985*. Marion Press, Battleford, Saskatchewan, 1985.

Clark, Herbert R. *Historical review of Saskatchewan Agricultural Societies Association*. University of Saskatchewan Printing Services, 1980.

Claus, Louis. *The Plate: A Royal Tradition*. Deneau Publishers & the Ontario Jockey Club, Toronto, 1984.

Coates, Ken and Fred McGuinness. *Pride of the Land: An Affectionate History of Brandon's Agricultural Exhibitions*. Pegus Publishers, Winnipeg, 1985.

Colin, Russell. *Saint John's Exhibitions*.

Dietz, Linda and Shirley Olekso. *Memories of Summers Past, 110th Anniversary of the Saskatoon Exhibition*. Saskatoon Heritage Society, 1996.

___. *History of the Saanich Fair*. Peninsula Printing Company, Sydney B.C., 1968.

Falhman, Jeanne. *Circling This Century: The Weyburn Agricultural Society*.

Fish, Charles. *Blue Ribbons and Burlesque: A Book of Country Fairs*. Countryman Press, Woodstock, Vermont, 1998.

___. *Conklin Shows Yearbook 1987*.

Fleury, Kay. *Let's go to the Fair: A History of the Weyburn Agricultural Society 1908-1968*.

Fowke, Vernon C. *Canadian Agricultural Policy: This Historical Pattern*. University of Toronto Press, 1978.

Frith, Joan. *Treatise of a Society: Prince Albert Exhibition Association*. Friesen Printers, Altona, Manitoba, 1983.

Grey, James H. *A Brand of Its Own: A History of the Calgary Stampede*.

Jones, David C., *Midways, Judges and Smooth-Tongued Fakirs*. Western Producer Prairie Books, Saskatoon, 1983.

Jones, Elwood. *Winners: 150 Years of the Peterborough Exhibition*. Peterborough

Agricultural Society, Peterborough, 1995.

Jones, Robert L. *A History of Agriculture in Ontario 1615-1880*. University of Toronto Press, 1946.

MacDonald, Cheryl. *Splendor in the Fall: Norfolk County Fair an Unbroken Heritage*. Norfolk County Agricultural Society, 1990.

MacEwan, Grant. *Agriculture on Parade*. Thomas Nelson, Toronto, 1950.

___. *Between the Red and The Rockies*. Univ. of Toronto Press, 1952.

Marti, Donald B., *Historical Directory of American Agricultural Fairs*. Greenwood Press, Westport Conn., 1986.

McCarry, John. *County Fairs: Where America Meets*. National Geographic Society, 1997.

McKennon, Joe. *A Pictorial History of the American Carnival*. Carnival Publisher of Sarasota, Fla., 1972.

Morley, Henry. *Memoirs of Bartholomews Fair*. Chatto & Windus, London, 1880.

Needles, Dan. *The Royal Agricultural Winter Fair, An Illustrated History*. Random House of Canada, Toronto, 1997.

Nelson, Derek. *The American State Fair*. MBI Publishing, Oscala, Wisconsin, 1999.

Posterman, Leslie. *Ordinary Like, Festival Days. Aesthetics in the Midwestern County Fair*. Smithsonian Institute, Washington, D.C., 1995.

Rasmussen, Carl. *Trail Blazers of Canadian Agriculture*. Agricultural Institute of Canada Foundation, Ottawa, 1995.

Reamore, Elmore G. *A History of Agriculture in Ontario, Volume One*. Hazel, Watson and Viney Ltd., Aylesbury, England, 1970.

Reynolds, Nila. *In Quest of Yesterday: A History of Haliburton County*. John Deyell Ltd., Lindsay, 1973.

Sanmiya, Inge. *A Celebration of Excellence. The History of the Western Fair Association*. Western Fair Association, London, (Ont), 2000.

Scott, Guy. *A History of Agricultural Societies and Fairs in Ontario, 1792-1992*. John Deyell Ltd., Lindsay, 1992.

Stevenson, Bonnie. *Agribition*. Centax Books, Regina, 1990.

Underwood, Amber. *Breaking Ground: The Story of the Ontario Ploughman's Association*. Ampersand Printing, 1987.

Walford, Cornelius. *Fairs Past and Present: A Chapter in the History of Commerce*. A.M. Kelley, New York, 1968.

Wilcox, George. *A History of Exhibitions and Stampedes in Medicine Hat*. Friesen, Altona, Manitoba, 1996.

___. *Once Upon a Century: 100 Year History of the C.N.E.* Robinson Publishing, Toronto, 1978.

Index

A

Aboriginals, *7*, 108
agricultural fairs, 2, 4, 8
    in the United States, 31
    see also fairs; school fairs
agricultural revolution, 28-29
agricultural societies, 3-4, 5, 7, 29, 30-31, 45, 198
    in Atlantic Canada, 37-39
    in Ontario, 41-46
    in Quebec, 40-41
    Western Canada, 46-52
airplanes, 111, 112-113
airships, *110*
Alberta fairs, 49-52
ancient fairs, 20
animal acts, 74-75, *103*
Armstrong Fair (BC), *92*, *130*
arts and crafts, 141
Atlantic Canada, 36-39
attendance, 3
Australia, 32-33
automobiles, 2, 6, 19, *98*

B

baby shows, 100-101
Baker, Jean, 182-184
bankruptcy, 85-86
barnstormers, 112-113
Barnum, P.T., 59-60, *175*
Bates, Thomas, 29
Battle River Fair and Stampede (AB), *18*
Baum, L. Frank, 88-89
Bayfield Fair (ON), 2, *181*
Belleville Fair (ON), *156*
Berkshire fairs, 29-30
Bishop, Tom, 118-119
boards of directors, 7, 136-137
Bobcaygeon Fair (ON), 168, *173*
Bonnie and Clyde, 76
Bostwick, Frank, 60-61
Bostwick-Frerari Carnival, 61, 67
Boyd and Lindeman's Mighty Midway, 66
Boyd, Mossom Martin, 168-169
Boys and Girls Clubs, 185-186
Brampton Fair (ON), *127*
Brandon Fair (MB), *47*, 48, *115*
Bridgewater Fair (NS), *159*
Brigden Fair (ON), *35*
British Columbia fairs, 14-16, 52-54, 56, 70-72
Bruce County school fair (ON), *180*
buffalo, *98*, 168
Bulkley Valley Fair (BC), *16*
burlesque, *81*, 82-84

C

Calgary Exhibition and Stampede, *7*, *174*
Calgary Fair, 103, *106*, *110*, *173*, *176*, *189*
Calgary Stampede, 51, 84, 94-97, 99, *172*, *174*
The Calgary Stampede (movie), 95-96
Canada
    fairs, 58, 36-37, 54
    international events, 120-121
    settlers, 6-7, 29, 120-121, *136*, *171*
Canadian Association of Fairs and Exhibitions (CAFE), 2, 3, 16
Canadian National Exhibition (CNE), 3, *8*, *9*, *75*, 85, 99, 108
    agricultural displays, *121-124*, *138*
    air show, *112*
    Crystal Palace, *143*
    grandstand shows, 109
    marathon swims, *100*
    Warrior's Day Parade, *172*
    and World War I, 128-129
Canadian Pacific Railway (CPR), *120*, 121, 145, *176*
Cannington Fair (SK), 49
Cardston Fair (AB), *176*
carnivals *see* midway
carousels *see* merry-go-rounds
Case, Jerome, 31
cattle, 153, *155*, *160*, 161, 167-169
Central Canada Exhibition (Ottawa), 85, *143*, *187*
chariot races, *96*, 97
Chicago World's Fair, 60-61, 68, 82, 87-89
chuckwagon races, 97
circuses, 59-60, *175*
Cody, Buffalo Bill, 16, 94
Coke, Thomas, 28
Coldstream Guards, *109*
Columbian World's Fair *see* Chicago World's Fair
commercial exhibits, 120-121
    displays, *121-127*
Community Centres Act (ON), 98-99
Conklin and Garrett Shows, 63, *66*, *74*
Conklin, Patty, 67, 84-85, 131
Conklin's Midway, *82*
contests, 100-101
    pulling, 158-159
    World Pumpkin Contest, 140
conventions, 13-16

Crystal Palace Exhibition, 87, 142
crystal palaces, *54*, *94*, 142-144

D

Defoe, Daniel, 26
demographics, 5-6
demolition derbies, *115*
Department of Agriculture, 4
depression *see* Great Depression
displays
    agricultural, *12-124*, *139*
    products, *125-127*
domestic arts, 132-133, 135-137
    see also homecrafts
Dominion Exhibition, Calgary, *10*, *144*, *175*, *177*
Dominion Exhibitions, 56
Dysart (Haliburton) Fair (ON), 44

E

Eaton's *see* T. Eaton Company
Edmonton Fair (AB), *58*, *61*, *68*, *69*, *70*, 109, *125*
elephants, *105*
English fairs, 23-27, 28
entertainment, 13-14, 31, 90, 109-110, 114-115
Erin Fair (ON), *17*
exhibit halls, 132-133
    arts and crafts, 141
    domestic arts, 135-137
    grain and field crops, 138-140
    judging exhibits, 133-135, 139-140
    and professional exhibitors, 137-138
Expo '67, 87

F

fairgrounds, 4, 17-19
fairs, 1
    appeal of, 10-12
    attendance, 3
    and children, 180, 187-188, 195
    and community, 5
    and education, 196-198
    and history, 5-8, 20-23
    memories of, 188-195
    names of, 8
    organizing, 13-16
    seasons, 2-3, 14, 178-179
    small vs. large, 116-118
    subcultures, 9
    women's role, 135-137
fakirs, 77
farm camps, 196-198
Ferris wheel, *30*, *59*, *68*, *88*, *89*
Ferris, William, 89
field crops *see* grain and field crops
Fife, David, 56-57

Finnish Giant, 73
fireworks, 113
flower shows, 140
4-H clubs, 185, *192*
Frakes, Frank, 113
freak shows, 7-74, *85*
Fredericton Fair (NB), *39*, *127*, *155*
Frerari Brothers, 6061

G

Galbraith, A.W., 152
gambling, 76-79, 92-94
games of chance *see* gaming concessions
gaming concessions, *71*, 76-79, 81-82
gilly cars, *60*, *63*
girlie shows, 82-84
Gooding Shows, 84
government grants, 149-150, 184, 198, 199
grain and field crops, 138-140
Grand Prairie Fair (AB), *135*
grandstand shows, 102-107, 109-110, 114-115
Granere, Dick, 113
Gray, Duncan, *188*
Great Depression, 84-85, 198-199

H

Haliburton Fair (ON), 44
Halton County Fair (ON), 93
Hand Hills Stampede (AB), *6*
Hays Classifications, 154
history, Canadian fairs, 5-8
hockey, 99
homecrafts, 4, 132-135
    see also domestic arts
horses, *95*, *96*, *103*, *151*, *160*, *163*
    diving, *12*, 76
    heavy-horse class, *149*, *150*, *151*, 159
    horse shows, 157-159
    in parades, 170
    racing, 90-94
    stampede events, 97
    and World War I, 129
horticulture, 140, *141*
hot-air balloons, 110, 112
Hudson's Bay Company, *13*, 121

I

Ilderton Fair (ON), *2*
immigrants *see* settlers
industrial exhibitions, 5
Ingeveld, Maurice, *10*
International Association of Fairs and Exhibitions
    (IAFE), 16
international events, 120-121

J
Jackson, Levi, 155
Jones, Johnny J., Midway, *58, 59, 60-63*

K
Kilmarnock Fair (MB), 152
King George III, 28
King's Plate, 92
Kinmount Fair (ON), *43, 93,* 112
Kleansol soap gimmick, 80

L
Lachute Fair (QC), 40-41
Lethbridge Fair (AB), 92, *94,* 130
"Little Egypt," 82, 89
livestock, 4, 28-29, 42, 146, 166
    breeding, 162, 168-169
    grooming, 155-156
    importing, 38
    judging, 150-155, 167
    show ring, 147-150
    small breeds, 161-163
    see also cattle; horses; llamas; sheep
llamas, 164-165
London Fair (ON), *137*

M
MacEwan, Grant, 188-195
machinery exhibits, *11, 126*
Manitoba fairs, 14, 46-48, 56
Marsh, Sadie, 81-82
McCall, Fred R., *111*
McGuinness, Fred, 13, 105-107
McGuinness, Wilson, *92*
McKennon, Joe, 58
mechanical rides, 59, 6769, 84
    see also Ferris wheel; merry-go-rounds; Roller-Boller
Medicine Hat Fair (AB), 79, 94, 187-188
medicine shows, *79*
Melfort Fair (SK), *136*
merry-go-rounds, *27,* 59, 67-68, *69*
Middle Ages, 20-23
midway, 58, 64, 66-67, 84
    accidents, 85-86
    Chicago World's Fair, 89
    and children, 81-82
    companies, 61-64
    con artists, 77-81
    history of, 58-61
    setbacks, 84-86
    vaudeville shows, 64-65
Miramichi Fair (NB), 7
Mission Fair (BC), *54*
motion pictures, 76
Muldrew, Jean, 134

musical performances, 109-110
Musical Ride (RCMP), *38*

N
Nelson Fair (BC), *132*
Neromus (bull wrestler), 104
New Westminster Fair (BC), *149*
New York State Fair, 31
Newfoundland fairs, 38-39
North Pacific Fairs Association, 14-16
North West Mounted Police, *38, 175*
Nottingham Goose Fair (England), 26-27
Nova Scotia fairs, 36-37

O
Ohio State Fair, 3
Oklahoma Wild West and Congress of Rough Riders Show, 97
Ontario Agricultural College, *152*
Ontario fairs, 14, 41-46, 55-56, 130, 154
Ontario Ploughman's Association, 35

P
Pacific National Exhibition (PNE), 19, 52, 54, 110, 114, *193*
pageants, 107-109
parades, 170
Peel County School Fair (ON), *166*
Penson, George, *8*
Perth County school fairs, 181-182
Peterborough Fair (ON), 97, 104-105, *129, 131*
petting zoos, 163-165
Picton Fair (ON), *142*
plowing matches, 33-35
Port Perry Fair (ON), *161*
Presley, Elvis, 114
Priddis and Millerville Fair (AB), *136*
Prince Albert Fair (SK), 92
Prince Andrew, 96
Prince of Wales, 95
Prince Rupert Fair (BC), *133, 137*
professional exhibitors, 137-138, 149
provincial fairs, 55-56
provincial organizations, 16
pulling contests, 158-159
pumpkin contests, 140

Q
Quebec fairs, 40-41
Queen's Plate, 92

R
Rand, Sally, 83
Ranfurly Fair (AB), 133
rating system, 2, 14, 64

Red Cross, *189*
Red Fife wheat, 56-57, 145
Regina Fair (SK), 198
regional differences, 11
Renfrew County Fair (ON), *134*
Rhedora, Jade, 83
rides *see* mechanical rides
rodeos, *53*, 94, 97
Roller-Boller ride, *80*
roller coasters, 68-69
Royal Melbourne Fair (Australia), 32-33
Royal North West Mounted Police Musical Ride, *38*
Royal Winter Fair, 167, 178
royalty, 95, 96

S
Saanich Fair (BC), 17
Saint John Exhibition (NB), 76
Saskatchewan fairs, 14, 48-49
Saskatoon Fair (SK), *47*, 112, *126*, 130, 157-158, 186, 198, 199
scams, 79-81
Scarborough Fair (ON), 5
school fairs, 141, *166*, 180-184, *185*
seasons for fairs, 2-3, 14, 178-179
seed fairs, 145
settlers, 6-7, 29, 120-121, *136*, *171*
sheep, 29
Sherbrooke Fair (QC), *40*, *148*, *160*, *161*
Showmen's League of America, 13-14, 16
Simcoe Fair (ON), 198
Simcoe, John Graves, 41
Simpsons Department Store, *177*
Skid Road, 70-72
Sousa, John Philip, 109
South Wellington Electoral Division Fair (ON), 18
sports, 98-99
spring fairs, 178-179
St. Bartholomew's Fair (England), 23-25
St. Louis Exposition, *121*
stampedes, 94, 97
strip shows *see* burlesque
stunt shows, 76, 112-113
Sturbridge Fair (England), 25-26
subcultures, 9
Sunderland Rural School Fair (ON), 182-184
swims, marathon, *100*

T
T. Eaton Company, *192*
theme parks, 86
Three Hills Fair (AB), *5*
Trelle, Herman, 145
Trounce, W.A., 135

U
United States fairs, 31
Upper Canada *see* Ontario fairs

V
Vancouver Fairgrounds, *129*
vaudeville shows, 64-65
vegetables *see* grain and field crops
Victoria Fair (BC), *54*
village theme, 9
volunteers, 9-10
Vulcan Fair (AB), *174*

W
Warnock, William, *140*
water shows, 75-76, *114*
Watson, Elkanah, 29, 162
Weadick, Guy, 95-96
weather, 14
see also seasons for fairs
Western Canada fairs, 46-52, 56, 108
Western Fair (ON), 156
Western Fairs Association, 14
Weyburn Fair (SK), *66*, 199
wheat, 56-57, 138, 145
Wheeler, Seeger, 145
Williamstown Fall Fair (ON), 183
Winnipeg Exhibition, *46*
Winnipeg Fair, 104
Winnipeg Stampede, *95*
Winston, D.C., 168
winter fairs, 178
women's exhibits, 135-137
Wood, Dorland, *187*
world events, 76
World Grain Show, 145
World of Mirth, 85
World War I, 128-129
World War II, 130-131
world's fairs, 27-28, 8-89

Y
Yorktown Fair (SK), *82*
Young, George, *100*
Yukon Horticultural and Industrial Fair, *53*

Z
Zacchini the Human Cannonball, *83*